An ELEMENTARY INTRODUCTION *to the* THEORY *of* PROBABILITY

By

B. V. GNEDENKO

and

A. Ya. KHINCHIN

AUTHORIZED EDITION

Translated from the Fifth Russian Edition by
LEO F. BORON
The Pennsylvania State University

with the editorial collaboration of
SIDNEY F. MACK
The Pennsylvania State University

D0109239

DOVER PUBLICATIONS, INC.
NEW YORK

This Dover edition, first published in 1962, is an authorized, unabridged translation of the fifth Russian edition published in 1961.

Manufactured in the United States by Courier Corporation
60155214
www.doverpublications.com

This translation is dedicated to the memory of
Prof. Dr.
ALEKSANDR YAKOVLEVICH KHINCHIN
who died on November 19, 1959,
leaving a superlative legacy of
mathematical works to future
generations

L. F. B.

FOREWORD TO THE FIFTH SOVIET EDITION

The present edition was prepared by me after the death of A. Ya. Khinchin, an eminent scientist and teacher. Many of the ideas and results in the modern development of the theory of probability are intimately connected with the name of Khinchin. The systematic utilization of the methods of set theory and the theory of functions of a real variable in the theory of probability, the construction of the foundations of the theory of stochastic processes, the extensive development of the theory of the summation of independent random variables, and also the construction of a new approach to the problems of statistical physics and the elegant system of its discussion—all this is due to Aleksandr Yakovlevich Khinchin. He shares with S. N. Bernshtein and A. N. Kolmogorov the honor of creating the Soviet school of probability theory, which plays an outstanding role in modern science. I consider myself fortunate to have been his student.

We wrote this booklet in the period of the victorious conclusion of the Great Patriotic War; this was naturally reflected in the elementary formulation of military problems which we used as examples. Now—fifteen years after the victory—in days when the entire country is covered with forests of new construction, it is natural to extend the subject matter in the examples to illustrate the general theoretical situation. It is for this reason therefore that, not changing the discussion and elementary character of the book, I have allowed myself the privilege of replacing a large number of examples by new ones. The same changes, with some negligible exceptions, were introduced by me also in the French edition of our booklet (Paris, 1960).

Moscow, October 6, 1960   B. V. GNEDENKO

ПРЕДИСЛОВИЕ К АМЕРИКАНСКОМУ ИЗДАНИЮ

За последние годы теория вероятностей приобрела исключительно большое значение как для развития самой математики, так и для прогресса буквально всех отраслей естествознания, техники и экономики. Теперь её роль начинает осознаваться в лингвистике и даже в археологии. Вот почему так важно как можно шире и разнообразнее популяризировать её идеи и результаты.

Во многих странах раздаются настойчивые голоса за введение элементов теории вероятностей в курс средней школы. Эту точку зрения разделял и покойны А. Я. Хинчин (1894–1959). Недавно мне удалось обнаружить небольшую его рукопись, в которой он изложил свои взгляды на место теории вероятностей в школьном преподавании математики и в очень общих чертах наметил объем и характер изложения.

Я счастлив, что настоящая книжка станет доступна американскому читателю. За те пятнадцать лет, которые протекли с момента выхода в свет первого русского издания, появилось много интересных работ, расширивших поле применений теории вероятностей и о которых можно увлекательно рассказать даже в популярной книжке. Однако, мне не хотелось бы нарушать ни план, ни стиль того, что было задумано моим учителем и мной в последние месяцы войны, пронесшейся ураганом по полям и городам моей Родины. Изменения коснулись лишь некоторых примеров, предметное содержание которых определялось временем написания книжки. Эти изменения внесены мной в пятое русское издание, которое должно выйти в свет почти одновременно с американским.

24.4.61   Б. В. Гнеденко

FOREWORD TO THE AMERICAN EDITION

In recent years, the theory of probability has acquired exceptionally great importance for the development of mathematics itself as well as for the progress of literally all branches of natural science, technology and economy. Its role is now beginning to be acknowledged in linguistics and even in archaeology. It is for this reason that it is essential to popularize its ideas and results as widely as possible and in all their varieties.

In many countries there is a persistent demand for the introduction of the elements of the theory of probability in the high-school curriculum. This point of view was also shared by A. Ya. Khinchin (1894–1959). Not long ago, I found a short manuscript of his in which he discussed his views on the place of the theory of probability in the teaching of school mathematics and he noted in general outline the content and nature of presentation.

I am happy that the present little book is accessible to the American reader. During the fifteen years that passed from the time the first Soviet edition was published, many interesting works appeared which extended the field of application of probability theory and about which one could tell in a captivating manner even in a popular booklet. However, I did not wish to disturb the plan or style of what was thought out by my teacher and myself in the last months of the war, which swept over the countryside and cities of my country like a hurricane. Changes touched upon only certain examples whose subject matter was determined by the time when the booklet was written. These changes were made by me in the fifth Soviet edition which is to be published almost simultaneously with the American edition.

April 24, 1961   B. V. Gnedenko

FOREWORD TO THE FIRST SOVIET EDITION

Acquaintance with the theoretical foundations of a mathematical science always enables one to apply more knowledgeably and actively the results of this science in practice. Likewise, in the area of probability theory, the situation is such that a large number of leaders (and occasionally also rank and file workers) in the military, in industry, agricultural economy, economy, etc., whose mathematical training is very limited, must deal with the practical applications of this science.

The present little book has as its aim to acquaint, in the most accessible form, the workers of this group with the fundamental concepts of probability theory and the methods of probability calculations. This booklet is completely accessible to all those who have completed the 10-year secondary school [ages 7–17 in the USSR]; it is almost entirely accessible to those who have completed the 7-year school also [ages 7–14 in the USSR]. In almost all its parts, the book is constructed on the basis of concrete, practical examples; in the choice of these examples, however, we were guided primarily not by their practical reality but by the illustrative value for the mastery of the corresponding theoretical situations.

Moscow, January 7, 1945

CONTENTS

PART I. PROBABILITIES

PART II. RANDOM VARIABLES

PART I

PROBABILITIES

CHAPTER 1

THE PROBABILITY OF AN EVENT

§ 1. The concept of probability

When we say that under given conditions of firing a marksman has 92% success we mean that of 100 shots fired by him under certain well-defined conditions (e.g., the same target at a prescribed distance, the same firearm, and so on), there are approximately 92 successes (and hence about 8 failures) *on the average*. Of course, there will not be exactly 92 successful shots out of every 100; sometimes there will be 91 or 90 of them, sometimes there will be 93 or 94; at times the number of successes can even be noticeably less or noticeably greater than 92; but *on the average* after many repetitions of shots under the same conditions, this percentage of target hits will remain unchanged as long as with the passage of time no essential changes take place in the firing conditions. (Otherwise, for example, our marksman could increase his mastery, and thereby increase the average percentage of target hits to 95 or higher.) And experience shows that for such a marksman, the number of successful shots per hundred will be close to 92; those hundreds, in which, for example, this number is less than 88 or greater than 96, although these will be encountered, will occur comparatively rarely. The figure 92% which serves as an index of mastery of our marksman is usually very *stable*; i.e., the percentage of target hits in the majority of shots (under the same conditions) will be almost the same for a given marksman—deviating rather significantly from its average value only in rare, exceptional cases.

Let us consider still another example. It is observed in a certain factory that under given conditions on the average 1.6% of the manufactured articles do not satisfy the standard and are rejected. This means that in a collection, say, of 1000 articles which have not yet been subjected to inspection, there will be approximately 16 which are unusable. Sometimes, of course, the number of rejected articles will be somewhat greater, sometimes somewhat less, but on the average this number will be close to 16, and in the majority of collections of 1000 articles it will also be close to 16. It is understood that here also we assume that the conditions of production are invariant; i.e., the

3

organization of the technological process, equipment, raw materials, qualification of workers, and so on, remain the same.

Clearly, one could introduce any number of such examples.   In all these cases, we see that in *homogeneous, numerous* operations performed under prescribed conditions (repeated firings, the mass producing of articles, and so on), the percentage of a certain type of event which is important to us (hitting the target, the fact that articles do not meet a fixed standard, and so on) will almost always remain approximately unchanged, only in rare cases deviating somewhat significantly from some average figure.   One can therefore say that this average figure is a characteristic index of the given operation (under prescribed, strictly established conditions).   The percentage of target hits describes for us the mastery of the marksman, the percentage of rejects gives us an estimate of how much of the production is of good quality.   It is therefore self-evident that the knowledge of such indices is very important in the most diverse areas: in military operations, technology, economy, physics, chemistry, and other fields; for it enables us not only to estimate the outcome of mass phenomena which have already occurred but also to foresee the outcome of a mass operation in the future.

If, under given firing conditions, a marksman hits the target on the average 92 times out of 100 shots, we say that for this marksman and under these conditions the *probability of hitting the target* is 92% (or 92/100 or 0.92).   If, under given conditions, on the average of every 1000 finished articles in a certain factory there are 16 rejects, then we say that the *probability of manufacturing a reject* is 0.016 or 1.6% for the given production.

But in general what do we call the probability of an event in a given mass operation?   It is now not difficult to answer this question.   A mass operation always consists in the repetition of a large number of identical individual operations (e.g., firing—of individual shots, mass production—the manufacture of individual articles, and so on).   We are interested in a well-defined result of individual operations (hitting the target in a single shot, the fact that an individual article is nonstandard, and so forth), and above all in the number of such results in some mass operation (how many shots will hit the target, how many articles will be rejected, and so on).   The percentage or, in general, the fractional part of such "successful" results in a given mass operation will be called the *probability* of this result—this is of importance to us.   In the second example it would be more appropriate to say "unsuccessful" results.   However, in the theory of probability it is conventional to call those results which lead to the realization of the event which interests us in a problem "successful." In this connection, one must always have in view that the question of the probability of an event (result) has meaning only under precisely defined conditions in which our mass operation proceeds. Every essential variation of these conditions causes, as a rule, a change in the probability of the event under consideration.

If the mass operation is such that event A (for example, hitting the target) is observed on the average a times in b individual operations (shots), then the probability of the event A under the given conditions is $\frac{a}{b}$ $\left(\text{or } \frac{100a}{b}\%\right)$. We can therefore say that the *probability* of a "successful" result of an individual operation is the *ratio of the number of such "successful" results observed to the number of these individual operations* constituting the prescribed mass operation. It is self-evident that if the probability of some event equals a/b, then in every collection of b individual operations this event can possibly occur more than a times and less than a times—it is only *on the average* that it occurs approximately a times. And in the majority of many such collections of b operations the number of occurrences of event A will be close to a—*particularly, if* b *is a large number.*

EXAMPLE 1. During the first quarter of the year, in a certain city there were born:

$$145 \text{ boys and } 135 \text{ girls in January}$$
$$142 \text{ ,, ,, } 136 \text{ ,, ,, February}$$
$$152 \text{ ,, ,, } 140 \text{ ,, ,, March.}$$

What is the probability that a boy is born? The fractional part of boy births is:

$$\frac{145}{280} \approx 0.518 = 51.8\% \text{ in January}$$

$$\frac{142}{278} \approx 0.511 = 51.1\% \text{ in February}$$

$$\frac{152}{292} \approx 0.520 = 52.0\% \text{ in March.}$$

We see that the arithmetic average of the fractional parts for the individual months is close to the number $0.516 = 51.6\%$, so the probability sought, under the given conditions, is approximately 0.516 or 51.6%. This number is well known in demography (which is the science whose domain is the study of population dynamics); it appears that the fractional part of boy births under usual conditions will not deviate significantly from this number during various periods of time.

EXAMPLE 2. At the beginning of the last century there was discovered a remarkable phenomenon, which received the name Brownian movement (after the English botanist Brown who discovered it). This phenomenon is that very fine particles of matter suspended in a liquid are in chaotic motion which is executed without any visible causes. For a long time the reason for this apparently spontaneous motion could not be clarified, until the kinetic theory of gases gave a simple and complete explanation: the movement of particles suspended in a liquid results from the collision of molecules of the liquid against these particles. The kinetic theory of gases enables one to calculate the probability that in a given volume of liquid there will not be a single particle of suspended matter, the probability that there will be one, two, three, and so on, such particles. A number of experiments were carried out with the purpose of verifying the predications of the theory.

We present the results of 518 observations, made by the Swedish chemist Svedberg, of very fine particles of gold suspended in water. It was found that in the portion of space under observation, not a single particle was observed 112 times, 1 particle was observed 168 times, 2 particles 130 times, 3 particles 69 times, 4 particles 32 times, 5 particles 5 times, 6 particles once, and finally, 7 particles once. The fractional part of the observed number of particles equals

0 particles: $\frac{112}{518} \approx 0.216$ 4 particles: $\frac{32}{518} \approx 0.062$

1 particle: $\frac{168}{518} \approx 0.325$ 5 „ $\frac{5}{518} \approx 0.010$

2 particles: $\frac{130}{518} \approx 0.251$ 6 „ $\frac{1}{518} \approx 0.002$

3 „ $\frac{69}{518} \approx 0.133$ 7 „ $\frac{1}{518} \approx 0.002.$

The results of the observations, as it turned out, coincided very well with the theoretically predicted probabilities.

EXAMPLE 3. In a number of problems which are important in practice, it is essential to know how frequently certain letters of the

Russian alphabet can occur in a text. Thus, for example, it is irrational to stock up the same number of all letters in forming a typographical font, since certain letters in the text are encountered significantly more frequently than others. Therefore, one strives to have a larger number of the letters which are encountered more frequently. Investigations performed on literary texts led to an estimate of the frequency of occurrence of the letters in the Russian alphabet, including the spaces between letters, which is summarized in the following table[1] (set up in the order of decreasing relative frequency of occurrence).

Thus, the indicated investigations show that on the average of 1000 spaces and letters selected at random in a text, the letter "ф" will occur in two places, the letter "к" in twenty-eight places, the letter "о" in ninety places, and there will be spaces between letters in one hundred and seventy-five places. These data are sufficiently valuable information for forming stock fonts.

In recent years similar investigations, no longer restricted to the statistics of letters in Russian texts, are beginning to be used extensively for the explanation of the peculiarities of the Russian language, and also of the literary style of various authors.

Letter	space	о	е, ё	а	и	т	н
Relative frequency	0.175	0.090	0.072	0.062	0.062	0.053	0.053
Letter	с	р	в	л	к	м	д
Relative frequency	0.045	0.040	0.038	0.035	0.028	0.026	0.025
Letter	п	у	я	ы	з	ь, ъ	б
Relative frequency	0.023	0.021	0.018	0.016	0.016	0.014	0.014
Letter	г	ч	й	х	ж	ю	ш
Relative frequency	0.013	0.012	0.010	0.009	0.007	0.006	0.006
Letter	ц	щ	э	ф			
Relative frequency	0.004	0.003	0.002	0.002			

Similar data relative to telegraph communications can be used for the creation of the most economical telegraph codes which would allow one to transmit messages by means of a smaller number of signs and, therefore, more rapidly. It has become clear that the telegraph codes utilized now are not sufficiently economical.

---

[1] This little table was adapted by the first-named author from the extraordinarily popular booklet *Probability and Information* by A. M. Yaglom and I. M. Yaglom, 2nd ed., Fizmatgiz, 1960.

§ 2. Impossible and certain events

The probability of an event, obviously, is always a positive number or zero.   It cannot be greater than unity because in the fraction by which it is defined the numerator cannot be greater than the denominator, for the number of "successful" operations cannot be greater than the number of all operations undertaken.

We agree to denote the probability of the event A by $P(A)$.   Whatever this event is, we have

$$0 \le P(A) \le 1.$$

The larger $P(A)$ is, the more often the event A occurs.   For example, the greater the probability that a marksman hits the target, the more often does he have successful shots.   If the probability of an event is very small, then it occurs rarely; if $P(A) = 0$, then the event either never occurs or it occurs very rarely, so that in practice one can consider it to be *impossible*.   In contrast, if $P(A)$ is close to unity, then in the fraction by which this probability is expressed, the numerator is close to the denominator, i.e., the overwhelming majority of operations are "successful"; if $P(A) = 1$, then the event A occurs always or almost always, so that in practice one can assume it to be, as one says, "certain," i.e., one can assume that its occurrence is *certain*.   If $P(A) = 1/2$, then the event A occurs in approximately half of all cases; this means that "successful" operations are observed approximately as often as "unsuccessful" ones.   If $P(A) > 1/2$, then the event A occurs more frequently than it does not occur; for $P(A) < 1/2$, we have the reverse phenomenon.

How small must the probability of an event be before we can assume it to be, in practice, impossible?   It is impossible to give a general answer to this question because everything depends on how important the event is with which we are dealing.   Thus, 0.01 is a small number. If we have a supply of shells and 0.01 is the probability that a shell will not explode upon falling, then this means that approximately 1% of the shots will be ineffective.   One can reconcile oneself to this!   But if we have a parachute and the probability that in a jump it will not open is 0.01, then it is of course impossible to reconcile oneself with this under any circumstances, because this means that in one out of a hundred jumps the valuable life of a parachutist will be lost.   These examples show that in every individual problem we must establish in advance, on the basis of practical considerations, how small the probability of an event ought to be in order that we can consider it to be impossible and of insignificant consequence to the undertaking at hand.

§ 3. Problem

PROBLEM.   One marksman has 80% as his average of target hits and another (under the same firing conditions) has 70%.   Find the probability of destroying the target if both marksman shoot at it simultaneously.   The target is assumed to be destroyed if at least one of the two bullets hits it.

First method of solution.   We assume that 100 double shots are fired. The target will be destroyed by the first marksman in approximately 80 of them.   There remain about 20 shots in which this marksman misses.   Since the second marksman destroys the target on the average 70 times in 100 shots and hence 7 times in 10 shots, we can expect that in these 20 shots in which the first marksman misses, the second succeeds in destroying the target approximately 14 times.   Thus, in all 100 shots, the target turns out to be destroyed approximately $80 + 14 = 94$ times. The probability of destroying the target under the simultaneous fire of both marksmen is therefore equal to 94% or 0.94.

Second method of solution.   We again assume that 100 double shots are fired.   We have already seen that in this connection the first marksman has approximately 20 misses.   Since the second marksman has approximately 30 misses per hundred shots and hence 3 misses per ten shots, one can expect that among those 20 shots in which the first marksman misses, there will be approximately 6 in which the second will also miss.   In each of these 6 shots, the target will remain undestroyed and in each of the remaining 94 shots at least one of the marksmen will shoot successfully and hence the target will be destroyed.   We again arrive at the result that for a double firing the target will be destroyed in approximately 94 cases in 100; i.e., that the probability of destruction is 94% or 0.94.

The problem we considered is very simple.   But, nonetheless, it already leads us to a very important result: there are cases when it is useful to know how to find, knowing the probabilities of certain events, the probabilities of other, more complicated events.   In fact, there are very many cases like this not only in military operations but also in every science and in every practical activity where we encounter mass phenomena.

Of course, it would be very inconvenient to search for the particular method of solution for every new problem of this sort encountered.

Science always endeavors to form general rules, the knowledge of which would readily permit one to solve mechanically or almost mechanically individual problems which are similar to one another. In the area of mass phenomena, the science which takes upon itself the formulation of such general rules is called the *theory of probability*. The first principles of this science will be given in this book.

The theory of probability is one chapter of mathematical science, like arithmetic or geometry.   Therefore, its path is the path of precise reasoning, and formulas, tables, diagrams, and so on, serve as its tools.

CHAPTER 2

RULE FOR THE ADDITION OF PROBABILITIES

§ 4. Derivation of the rule for the addition of probabilities

The simplest and most important rule used in the calculation of probabilities is the *addition rule*, which we shall now consider.

In firing at a target, depicted in Fig. 1, for every marksman standing at a prescribed distance, there is a certain probability of hitting each of the regions 1, 2, 3, 4, 5, 6. Suppose that for some marksman the probability of hitting region 1 is 0.24 and that the probability of hitting region 2 is 0.17. As we already know, this means that of one hundred bullets shot by this marksman, 24 bullets (on the average) hit region 1 and 17 bullets hit region 2.

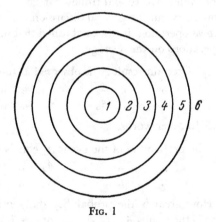

Fig. 1

Suppose that, in some competition, a shot is adjudged "exceptional" if the bullet falls into region 1 and "good" if it falls into region 2. What is the probability that the marksman's shot is either good or exceptional?

It is easy to answer this question. Of every hundred bullets shot by the marksman, approximately 24 fall into region 1 and approximately 17 into region 2. This means that of every hundred bullets there will be approximately $24+17=41$ which will fall into either region 1 or into region 2. The probability sought therefore equals $0.41 = 0.24 + 0.17$. Consequently, *the probability that the shot will be either exceptional or good equals the sum of the probabilities of the exceptional and good shots.*

Let us consider still another example. A passenger is waiting for trolley No. 26 or No. 16 at a trolley stop at which trolleys with one of the four route Nos. 16, 22, 26, and 31 stop. Assuming that the trolleys of all routes appear on the average equally frequently, find the probability that the first trolley appearing at the stop will have the route needed by the passenger.

Clearly, the probability that trolley No. 16 will be the first to appear at the stop equals 1/4; the probability that trolley No. 26 will be the first is the same. So, the probability sought is obviously equal to 1/2. But

$$1/2 = 1/4 + 1/4;$$

therefore we can say that the probability that trolley No. 16 or trolley No. 26 will appear first equals the sum of the probabilities of the appearance of trolley No. 16 and trolley No. 26.

We can now carry out the general discussion. In the performance of a certain mass operation, it was established that in every series of b individual operations on the average

$$\text{a certain result } A_1 \text{ is observed } a_1 \text{ times}$$
$$\text{,, ,, } A_2 \text{ ,, } a_2 \text{ ,,}$$
$$\text{,, ,, } A_3 \text{ ,, } a_3 \text{ ,,}$$

and so forth. In other words,

$$\text{the probability of the event } A_1 \text{ equals } a_1/b$$
$$\text{,, ,, ,, } A_2 \text{ ,, } a_2/b$$
$$\text{,, ,, ,, } A_3 \text{ ,, } a_3/b$$

and so on. How great is the probability that, in some individual operation, one of the results $A_1, A_2, A_3, \ldots$ occurs, it being immaterial which one?

The event of interest can be called "A_1 or A_2 or A_3 or $\ldots$." (Here and in other similar cases the ellipsis dots [$\ldots$] denote "and so forth.") In a series of b operations, this event occurs $a_1 + a_2 + a_3 + \ldots$ times; this means that the probability sought equals

$$\frac{a_1 + a_2 + a_3 + \ldots}{b} = \frac{a_1}{b} + \frac{a_2}{b} + \frac{a_3}{b} + \ldots$$

which can be written as the following formula:

$$P(A_1 \text{ or } A_2 \text{ or } A_3 \text{ or } \ldots) = P(A_1) + P(A_2) + P(A_3) + \ldots.$$

In this connection, in our examples as well as in our general discussion, we always assume that any two of the results considered (for instance, A_1 and A_2) are *mutually incompatible,* i.e., they cannot be observed together in the same individual operation. For instance, the trolley arriving cannot simultaneously be from a needed and not-needed route—it either satisfies the requirement of the passenger or it does not. This assumption concerning the mutual incompatibility of the individual results is very important, for without it the addition rule becomes invalid and its application leads to serious errors. We consider, for example, the problem we solved at the end of the preceding section (see page 9). There we even found the probability that for a double shot either one or the other shot will hit the target, in which connection for the first marksman the probability of hitting the target equals 0.8 and for the second 0.7. If we wished to apply the addition rule to the solution of this problem, then we would at once have found that the probability sought equals $0.8 + 0.7 = 1.5$ which is manifestly absurd since we already know that the probability of an event cannot be greater than unity. We arrived at this invalid and meaningless answer because we applied the addition law to a case where one must not apply it: the two results we are dealing with in this problem are *mutually compatible,* inasmuch as it is entirely possible that both marksmen destroy the target with the same double shot. A significant portion of errors which novices make in the computation of probabilities is due in fact to such an invalid application of the addition rule. It is therefore necessary to guard carefully against this error and verify in every application of the addition rule whether in fact, among those events to which we wish to apply it, every pair is mutually incompatible.

We can now give a general formulation of the addition rule.

ADDITION RULE. *The probability of occurrence in a certain operation of any one of the results* $A_1, A_2, \ldots, A_n$ *(it being immaterial which one) is equal to the sum of the probabilities of these results, provided that every pair of them is mutually incompatible.*

§ 5. Complete system of events

In the Third (Soviet) Government Loan (TSGL) for the reconstruction and development of the national economy, in the course of the twenty-year period of its operation, a third of the bonds win and the remaining two-thirds are drawn in a lottery and are paid off at the nominal rate. In other words, for this loan each bond has a probability equal to 1/3 of winning and a probability equal to 2/3 of being drawn in a lottery. Winning and being drawn in a lottery are *complementary events*; i.e., they are two events such that one and only one of them must necessarily occur for every bond. The sum of their probabilities is

$$\frac{1}{3} + \frac{2}{3} = 1,$$

and this is not accidental. In general, if A_1 and A_2 are two complementary events and if in a series of b operations the event A_1 occurs a_1 times and the event A_2 occurs a_2 times, then, obviously, $a_1 + a_2 = b$. But

$$P(A_1) = \frac{a_1}{b}, \qquad P(A_2) = \frac{a_2}{b},$$

so that

$$P(A_1) + P(A_2) = \frac{a_1}{b} + \frac{a_2}{b} = \frac{a_1 + a_2}{b} = 1.$$

This same result can also be obtained from the addition rule: since complementary events are mutually incompatible, we have

$$P(A_1) + P(A_2) = P(A_1 \text{ or } A_2).$$

But the event "A_1 or A_2" is a certain event since it follows from the definition of complementary events that it certainly must occur; therefore, its probability equals unity and we again obtain

$$P(A_1) + P(A_2) = 1.$$

The sum of the probabilities of two complementary events equals unity.

This rule admits of a very important generalization which can be proved by the same method. Suppose we have n events $A_1, A_2, \ldots,$ A_n (where n is an arbitrary positive integer) such that in each individual operation one and only one of these events must necessarily occur; we agree to call such a group of events a *complete system*. In particular, every pair of complementary events, obviously, constitutes a complete system.

The sum of the probabilities of events constituting a complete system is equal to unity.

In fact, according to the definition of a complete system, any two events in this system are mutually incompatible, so that the addition rule yields

$$P(A_1) + P(A_2) + \ldots + P(A_n) = P(A_1 \text{ or } A_2 \text{ or } \ldots \text{ or } A_n).$$

But the right member of this equality is the probability of a certain event and it therefore equals unity; thus, for a complete system, we have

$$P(A_1) + P(A_2) + \ldots + P(A_n) = 1,$$

which was to be proved.

EXAMPLE 1.   Of every 100 target shots (target depicted in Fig. 1 on page 11), a marksman has on the average

44 hits in region 1
30 ,, ,, 2
15 ,, ,, 3
6 ,, ,, 4
4 ,, ,, 5
1 hit ,, 6

$(44+30+15+6+4+1 = 100)$.   These six firing results obviously constitute a complete system of events.   Their probabilities are equal to

$$0.44, \quad 0.30, \quad 0.15, \quad 0.06, \quad 0.04, \quad 0.01,$$

respectively; we have

$$0.44 + 0.30 + 0.15 + 0.06 + 0.04 + 0.01 = 1.$$

Shots falling completely or partially into the region 6 do not hit the target at all and cannot be considered; this does not, however, hinder finding the probability of falling into this region, for which it is sufficient to subtract from unity the sum of the probabilities of falling into all the other regions.

EXAMPLE 2.   Statistics show that at a certain weaving factory, of every hundred stoppages of a weaving machine requiring the subsequent work of the weaver, on the average,

22 occur due to a break in the warp thread
31 ,, ,, ,, ,, woof ,,
27 ,, ,, change in the shuttle
3 occur due to breakage of the shuttlecock

and the remaining stoppages of the machine are due to other reasons.

We see that besides other reasons for the stoppage of the machine, there are four definite reasons whose probabilities are equal to

$$0.22, \quad 0.31, \quad 0.27, \quad 0.03,$$

respectively.   The sum of these probabilities equals 0.83.   Together with the other reasons, the reasons pointed out for stoppage of the machine constitute a complete system of events; therefore, the probability of stoppage of the machine from other causes equals

$$1 - 0.83 = 0.17.$$

§ 6. Examples

We frequently successfully base the so-called a priori, i.e., pretrial, calculation of probabilities on the theorem concerning a complete system of events which we have established.   Suppose, for example, that we are studying the falling of cosmic particles into a

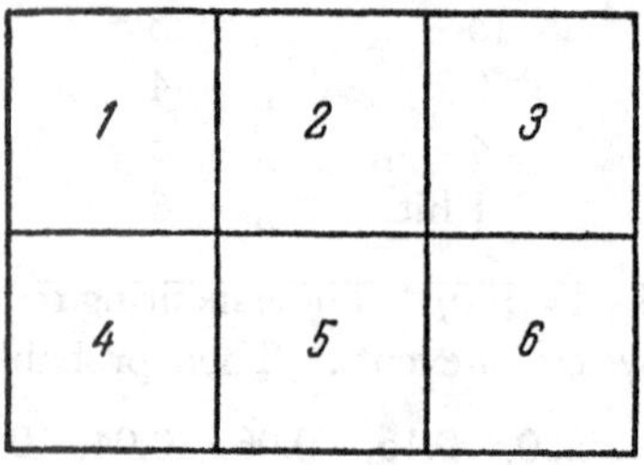

| 1 | 2 | 3 |
| 4 | 5 | 6 |

Fig. 2

small area of rectangular form (see Fig. 2)—this area being subdivided into the 6 equal squares numbered in the figure.   The subareas of interest find themselves under the same conditions and therefore there is no basis for assuming that particles will fall into any one of these six squares more often than another.   We therefore assume that on the average particles will fall into each of the six squares equally frequently, i.e., that the probabilities $p_1, p_2, p_3, p_4, p_5, p_6$ of falling into these squares are equal.   If we assume that we are interested only in particles which fall into this area, then it will follow from this that each of the numbers p equals $1/6$, inasmuch as these numbers are equal and their sum equals unity by virtue of the theorem we proved above.   Of course this result, which is based on a number of assumptions, requires experimental verification for its affirmation.   We have, however, become so accustomed in such cases to obtaining excellent agreement between our theoretical assumptions and their experimental verifications that we can depend on the theoretically deduced probabilities for all practical purposes. We usually say in such cases that the given operation can have n distinct, mutually *equi-probable* results (thus, in our example of cosmic particles falling into an area, depicted in Fig. 2, the result is that the particle falls into one of the six squares). The probability of each of these n results is equal in this case to $1/n$. The importance of this type of a priori reasoning is that in many cases it enables us to foresee the probability of an event under conditions where its determination by repetitive operations is either absolutely impossible or extremely difficult.

EXAMPLE 1.   In the case of government loan bonds, the numbers of a series are usually expressed by five-digit numbers. Suppose we wish to find the probability that the last digit, taken at random from a winning series, equals 7 (as, for example, in the series No. 59607). In accordance with our definition of probability, we ought to consider, for this purpose, a long series of lottery tables and calculate how many winning series have numbers ending in the digit 7; the ratio of this number to the total number of winning series will then be the probability sought. However, we have every reason to assume that any one of the ten digits 0, 1, 2, 3, 4, 5, 6, 7, 8, 9 has as much of a chance to appear in the last place in a number of the winning series as any other. Therefore, without any hesitation, we make the assumption that the probability sought equals 0.1. The reader can easily verify the legitimacy of this theoretical "foresight": carry out all necessary calculations within the framework of any one lottery table and verify that in reality each of the 10 digits will appear in the last place in approximately $1/10$ of all cases.

EXAMPLE 2.   A telephone line connecting two points A and B at a distance of 2 km. broke at an unknown spot. What is the probability that it broke no farther than 450 m. from the point A? Mentally subdividing the entire line into individual meters, we can assume, by virtue of the actual homogeneity of all these parts, that the probability of breakage is the same for every meter. From this, similar to the preceding, we easily find that the required probability equals

$$\frac{450}{2000} = 0.225.$$

CHAPTER 3

CONDITIONAL PROBABILITIES AND THE MULTIPLICATION RULE

§ 7. The concept of conditional probability

Electric light bulbs are manufactured at two plants—the first plant furnishes 70% and the second 30% of all required production of bulbs. At the first plant, among every 100 bulbs 83 are on the average standard,[1] whereas only 63 per hundred are standard at the second plant.

It can easily be computed from these data that on the average each set of 100 electric light bulbs purchased by a consumer will contain 77 standard bulbs and, consequently, the probability of buying a standard bulb equals 0.77.[2] But we shall now assume that we have made it clear that the bulbs on stock in a store were manufactured at the first plant. Then the probability that the bulb is standard will change—it will equal $83/100 = 0.83$.

The example just considered shows that the addition to the general conditions under which an operation takes place (in our case this is the purchase of the bulbs) of some essentially new condition (in our example this is knowledge of the fact that the bulb was produced by one or the other of the plants) can change the probability of some result of an individual operation. But this is understandable; for the very definition of the concept of probability requires that the totality of conditions under which a given mass operation occurs be precisely defined. By adding any new condition to this collection of conditions we, generally speaking, change this collection in an essential way. Our mass operation takes place after this addition under new conditions; in reality, this is already another operation, and therefore the probability of some result in it will no longer be the same as that under the initial conditions.

We thus have two distinct probabilities of the same event—i.e., the

---

[1] In this regard, we call a bulb "standard" (i.e., it meets certain *standard* requirements) if it is capable of functioning no less than 1200 hours; otherwise, the bulb will be called substandard.

[2] In fact, we have $0.83 \cdot 70 + 0.63 \cdot 30 = 77$.

purchase of a standard bulb—but these probabilities are calculated under different conditions. As long as we do not set down an additional condition (e.g., not considering where the bulb was manufactured), we take the *unconditional probability* of purchasing a standard bulb as equal to 0.77; but upon placing an additional condition (that the bulb was manufactured in the first plant) we obtain the *conditional probability* 0.83, which differs somewhat from the preceding. If we denote by A the event of purchasing a standard bulb and by B the event that it was manufactured in the first plant, then we usually denote by $P(A)$ the unconditional probability of event A and by $P_B(A)$ the probability of the same event under the condition that event B has occurred, i.e., that the bulb was manufactured by the first plant. We thus have $P(A) = 0.77$, $P_B(A) = 0.83$.

Since one can discuss the probability of a result of a given operation only under certain precisely defined conditions, every probability is, strictly speaking, a conditional probability; unconditional probabilities cannot exist in the literal sense of this word. In the majority of concrete problems, however, the situation is such that at the basis of all operations considered in a given problem there lies some well-defined set of conditions K which are assumed satisfied for all operations. If in the calculation of some probability no other conditions except the set K are assumed, then we shall call such a probability *unconditional*; the probability calculated under the assumption that further precisely prescribed conditions, besides the set of conditions K common to all operations, are satisfied will be called *conditional*.

Thus, in our example, we assume, of course, that the manufacture of a bulb occurs under certain well-defined conditions which remain the same for all bulbs which are placed on sale. This assumption is so unavoidable and self-evident that in the formulation of problems we did not even find it necessary to mention it. If we do not place any additional conditions on the given bulb, then the probability of some result in the testing of the bulb will be called unconditional. But if, over and above these conditions, we make still other, additional requirements, then the probabilities computed under these requirements will now be conditional.

EXAMPLE 1.   In the problem we described at the beginning of the present section, the probability that the bulb was manufactured by the second plant obviously equals 0.3. It is established that the bulb is of standard quality. After this observation, what is the probability that this bulb was manufactured at the second plant?

Among every 1000 bulbs put on the market, on the average 770 bulbs are of standard quality—and of this number 581 bulbs came from the first plant and 189 bulbs came from the second.[1] After making this observation, the probability of issuing a bulb by the second plant therefore becomes $189/770 \approx 0.245$. This is the conditional probability of issuing a bulb by the second plant, calculated under the assumption that the given bulb is standard. Using our previous notation, we can write $P(\overline{B}) = 0.3$ and $P_A(\overline{B}) \approx 0.245$, where the event $\overline{B}$ denotes the nonoccurrence of the event B.

EXAMPLE 2.   Observations over a period of many years carried out in a certain region showed that among 100,000 children who have attained the age of 9, on the average 82,277 live to 40 and 37,977 live to 70.   Find the probability that a person who attains the age 40 will also live to 70.

Since on the average 37,977 of the 82,277 forty-year-olds live to 70, the probability that a person aged 40 will live to 70 equals $37,977/82,277 \approx 0.46$.

If we denote by A the first event (that a nine-year-old child lives to 70) and by B the second event (that this child attains the age 40), then obviously, we have $P(A) = 0.37,977 \approx 0.38$ and $P_B(A) \approx 0.46$.

§ 8. Derivation of the rule for the multiplication of probabilities

We now return to the first example in the preceding section. Among every 1000 bulbs placed on the market, on the average 300 were manufactured at the second plant, and among these 300 bulbs on the average 189 are of standard quality.   We deduce from this that the probability that the bulb was manufactured at the second plant (i.e., event $\overline{B}$) equals $P(\overline{B}) = 300/1000 = 0.3$ and the probability that it is of standard quality, under the condition that it was manufactured at the second plant, equals $P_{\overline{B}}(A) = 189/300 = 0.63$.

Since, out of every 1000 bulbs, 189 were manufactured at the second plant and are at the same time of standard quality, the probability of the simultaneous occurrence of the events A and $\overline{B}$ equals

$$P(A \text{ and } \overline{B}) = \frac{189}{1000} = \frac{300}{1000} \cdot \frac{189}{300} = P(\overline{B}) \cdot P_{\overline{B}}(A).$$

[1] This can easily be calculated as follows.   Among every 1000 bulbs, on the average 700 were manufactured at the first plant, and among every 100 bulbs from the first plant on the average 83 are of standard quality. Consequently, among 700 bulbs from the first plant, on the average $7 \cdot 83 = 581$ will be of standard quality.   The remaining 189 bulbs of standard quality were produced at the second plant.

This "multiplication rule" can also be easily extended to the general case.   Suppose in every sequence of n operations, the result B occurs on the average m times, and that in every sequence of m such operations in which the result B is observed, the result A occurs l times.   Then, in every sequence of n operations, the simultaneous occurrence of the events B and A will be observed on the average l times.   Thus,

$$P(B) = \frac{m}{n}, \qquad P_B(A) = \frac{l}{m},$$

$$P(A \text{ and } B) = \frac{l}{n} = \frac{m}{n} \cdot \frac{l}{m} = P(B) \cdot P_B(A). \tag{1}$$

MULTIPLICATION RULE.   *The probability of the simultaneous occurrence of two events equals the product of the probability of the first event with the conditional probability of the second, computed under the assumption that the first event has occurred.*

It is understood that we can call either of the two given events the first so that on an equal basis with formula (1) we can also write

$$P(A \text{ and } B) = P(A) \cdot P_A(B), \tag{1'}$$

from which we obtain the important relation:

$$P(A) \cdot P_A(B) = P(B) \cdot P_B(A). \tag{2}$$

In our example, we had

$$P(A \text{ and } \overline{B}) = \frac{189}{1000}, \qquad P(A) = \frac{77}{100}, \qquad P_A(\overline{B}) = \frac{189}{770};$$

and this shows that formula (1') is satisfied.

EXAMPLE.   At a certain enterprise, 96% of the articles are judged to be usable (event A); out of every hundred usable articles, on the average 75 turn out to be of the first sort (event B).   Find the probability that an article manufactured at this enterprise is of the first sort.

We seek $P(A \text{ and } B)$ since, in order that an article be of the first sort, it is necessary that it be usable (event A) and of the first sort (event B).

By virtue of the conditions of the problem, $P(A) = 0.96$ and $P_A(B) = 0.75$.   Therefore, on the basis of formula (1'), $P(A \text{ and } B) = 0.96 \cdot 0.75 = 0.72$.

§ 9. Independent events

Two skeins of yarn, manufactured on different machines, were tested for strength.   It turned out that a sample of prescribed length taken from the first skein held a definite standard load with probability 0.84 and that from the second skein with probability 0.78.[1]   Find the probability that two samples of yarn, taken from two different skeins, are both capable of supporting the standard load.

We denote by A the event that the sample taken from the first skein supports the standard load and by B the analogous event for the sample from the second skein.   Since we are seeking $P(A \text{ and } B)$, we apply the multiplication rule:

$$P(A \text{ and } B) = P(A) \cdot P_A(B).$$

Here we obviously have $P(A) = 0.84$; but what is $P_A(B)$?   According to the general definition of conditional probabilities, this is the probability that the sample of yarn from the second skein will support the standard load if the sample from the first skein supported such a load.   But the probability of event B does not depend on whether or not event A has occurred, for these tests can be carried out simultaneously and the yarn samples are chosen from completely unrelated skeins, manufactured on different machines.   In practice, this means that the percentage of trials in which the yarn from the second skein supports the standard load does not depend on the strength of the sample from the first skein; i.e.,

$$P_A(B) = P(B) = 0.78.$$

It follows from this that

$$P(A \text{ and } B) = P(A) \cdot P(B) = 0.84 \cdot 0.78 = 0.6552.$$

The peculiarity which distinguishes this example from the preceding ones consists, as we see, in that here the probability of the result B is not changed by the fact that to the general conditions we add the requirement that the event A occur.   In other words, the conditional probability $P_A(B)$ equals the unconditional probability $P(B)$.   In this case we will say, briefly, that *the event* B *does not depend on the event* A.

It can easily be verified that if B does not depend on A, then A also does not depend on B.   In fact, if $P_A(B) = P(B)$, then by virtue of formula (2) $P_B(A) = P(A)$ and this means that the event A does not depend on the event B.   Thus, the independence of two events is a *mutual* (or *dual*) property.   We see that for mutually independent events, the multiplication rule has a particularly simple form:

$$P(A \text{ and } B) = P(A) \cdot P(B). \tag{3}$$

---

[1] If the standard load equals, say, 400 grams, then this means the following: among 100 samples taken from the first skein, 84 samples on the average support such a load and 16 do not support it and break.

As in every application of the addition rule it is necessary to establish in advance the mutual incompatibility of the given events, so in every application of rule (3) it is necessary to verify that the events A and B are mutually independent.   Disregard for these instructions leads to errors.   If the events A and B are mutually dependent, then formula (3) is not valid and must be replaced by the more general formula (1) or 1').

Rule (3) is easily generalized to the case of seeking the probability of the occurrence of not two, but of three or more mutually independent events.   Suppose, for example, that we have three mutually independent events A, B, C (this means that the probability of any one of them does not depend on the occurrence or the nonoccurrence of the other two events).   Since the events A, B and C are mutually independent, we have, by rule (3):

$$P(A \text{ and } B \text{ and } C) = P(A \text{ and } B) \cdot P(C).$$

Now if we substitute here for $P(A \text{ and } B)$ the expression for this probability from formula (3), we find:

$$P(A \text{ and } B \text{ and } C) = P(A) \cdot P(B) \cdot P(C). \tag{4}$$

Clearly, such a rule holds in the case when the set under consideration contains an arbitrary number of events as long as these events are mutually independent (i.e., the probability of each of them does not depend on the occurrence or nonoccurrence of the remaining events).

The probability of the simultaneous occurrence of any number of mutually independent events equals the product of the probabilities of these events.

EXAMPLE 1.   A worker operates three machines.   The probability that for the duration of one hour a machine does not require the attention of the worker equals 0.9 for the first machine, 0.8 for the second, and 0.85 for the third.   Find the probability that for the duration of an hour none of the machines requires the worker's attention.

Assuming that the machines work independently of each other, we find, by formula (4), that the probability sought is

$$0.9 \cdot 0.8 \cdot 0.85 = 0.612.$$

EXAMPLE 2.   Under the conditions of Example 1, find the probability that at least one of the three machines does not require the attention of the worker for the duration of one hour.

In this problem, we are dealing with a probability of the form $P(A \text{ or } B \text{ or } C)$ and, therefore, we of course think first of all of the addition rule.   However, we soon realize that this rule is not applicable in the present case inasmuch as any two of the three events considered can occur simultaneously; for nothing hinders any two machines from working without being given attention for the duration of the same hour. Moreover, independently of this line of reasoning, we at once see that the sum of the three given probabilities is significantly larger than unity and hence we cannot compute the probability in this way.

To solve the problem as stated, we note that the probability that a machine requires the attention of the worker equals 0.1 for the first machine, 0.2 for the second, and 0.15 for the third. Since these three events are mutually independent, the probability that all these events are realized equals

$$0.1 \cdot 0.2 \cdot 0.15 = 0.0003,$$

according to rule (4). But the events "all three machines require attention" and "at least one of the three machines operates without receiving attention" clearly represent a pair of complementary events. Therefore, the sum of their probabilities equals unity and, consequently, the probability sought equals $1 - 0.0003 = 0.9997$. When the probability of an event is as close to unity as this, then this event can in practice be assumed to be certain. This means that almost always, in the course of an hour, at least one of the three machines will operate without receiving attention.

EXAMPLE 3.   Under certain definite conditions, the probability of destroying an enemy's plane with a rifle shot equals 0.004. Find the probability of destroying an enemy plane when 250 rifles are fired simultaneously.

For each shot, the probability is $1 - 0.004 = 0.996$ that the plane will not be downed. The probability that it will not be downed by all 250 shots equals, according to the multiplication rule for independent events, the product of 250 factors each of which equals 0.996, i.e., it is equal to $(0.996)^{250}$. And the probability that at least one of the 250 shots proves to be sufficient for downing the plane is therefore equal to

$$1 - (0.996)^{250}.$$

A detailed calculation, which will not be carried out here, shows that this number is approximately equal to 5/8. Thus, although the probability of downing an enemy plane by one rifle shot is negligibly small—0.004—with the simultaneous firing from a large number of rifles, the probability of the desired result becomes very significant.

The line of reasoning which we utilized in the last two examples can easily be generalized and leads to an important general rule.   In both cases, we were dealing with the probability $P(A_1$ or A_2 or A_3 ... or $A_n)$ of the occurrence of at least one of several mutually independent events $A_1, A_2, \ldots, A_n$.   If we denote by $\bar{A}_k$ the event that A_k will not occur, then the events A_k and $\bar{A}_k$ are complementary, so that

$$P(A_k) + P(\bar{A}_k) = 1.$$

On the other hand, the events $\bar{A}_1, \bar{A}_2, \ldots, \bar{A}_n$ are obviously mutually independent so that

$$\begin{aligned}
P(\bar{A}_1 \text{ and } \bar{A}_2 \text{ and } \ldots \text{ and } \bar{A}_n) &= P(\bar{A}_1) \cdot P(\bar{A}_2) \ldots P(\bar{A}_n) \\
&= [1 - P(A_1)] \cdot [1 - P(A_2)] \ldots \\
&\quad [1 - P(A_n)].
\end{aligned}$$

Finally, the events $(A_1$ or A_2 or ... or $A_n)$ and $(\bar{A}_1$ and $\bar{A}_2$ and ... and $\bar{A}_n)$ obviously are complementary; that is, one of the following: either at least one of the events A_k occurs or all the events $\bar{A}_k$ occur. Therefore,

$$\begin{aligned}
P(A_1 \text{ or } A_2 \text{ or } \ldots \text{ or } A_n) &= 1 - P(\bar{A}_1 \text{ and } \bar{A}_2 \text{ and } \ldots \text{ and } \bar{A}_n) \\
&= 1 - [1 - P(A_1)] \cdot [1 - P(A_2)] \ldots [1 - P(A_n)]. \quad (5)
\end{aligned}$$

This important formula, which enables one to calculate the probability of the occurrence *of at least one* of the events $A_1, A_2, \ldots, A_n$ on the basis of the given probabilities of these events, is valid if, and only if, these events are mutually independent.   In the particular case when all the events A_k have the same probability p (as was the case in Example 3, above) we have:

$$P(A_1 \text{ or } A_2 \text{ or } \ldots \text{ or } A_n) = 1 - (1 - p)^n. \quad (6)$$

EXAMPLE 4.   An instrument part is being lathed in the form of a rectangular parallelepiped. The part is considered usable if the length of each of its edges deviates by no more than 0.01 mm. from prescribed dimensions. If the probability of deviations exceeding 0.01 mm. is

$$p_1 = 0.08 \text{ along the length of the parallelepiped}$$
$$p_2 = 0.12 \quad ,, \qquad \text{width} \quad ,, \qquad ,,$$
$$p_3 = 0.10 \quad ,, \qquad \text{height} \quad ,, \qquad ,,$$

find the probability P that the part is not usable.

For the part to be unusable, it is necessary that at least in one of the three directions the deviation from the prescribed dimension exceed

0.01 mm. Since these three events can usually be assumed mutually independent (because they are basically due to different causes), to solve the problem we can apply formula (5); this yields

$$P = 1 - (1-p_1) \cdot (1-p_2) \cdot (1-p_3) \approx 0.27.$$

Consequently, we can assume that of every 100 parts approximately 73 on the average turn out to be usable.

CHAPTER 4

CONSEQUENCES OF THE ADDITION AND MULTIPLICATION RULES

§ 10. Derivation of certain inequalities

We turn again to the electric light bulb example of the preceding chapter (see page 18). We introduce the following notation for events:

A—the bulb is of standard quality

$\bar{A}$— the bulb is of substandard quality

B—the bulb was manufactured at the first plant

$\bar{B}$—the bulb was manufactured at the second plant.

Obviously, events A and $\bar{A}$ constitute a pair of complementary events; the events B and $\bar{B}$ form a pair of the same sort.

If the bulb is of standard quality (A), then either it was manufactured by the first plant $(A$ and $B)$ or by the second $(A$ and $\bar{B})$. Since the last two events, evidently, are incompatible with one another, we have, according to the addition rule

$$P(A) = P(A \text{ and } B) + P(A \text{ and } \bar{B}). \qquad (1)$$

In the same way, we find that

$$P(B) = P(A \text{ and } B) + P(\bar{A} \text{ and } B). \qquad (2)$$

Finally, we consider the event $(A$ or $B)$; we obviously have the following three possibilities for its occurrence:

1) A and B, 2) A and $\bar{B}$, 3) $\bar{A}$ and B.

Of these three possibilities, any two are incompatible with one another; therefore, by the addition rule, we have

$$P(A \text{ or } B) = P(A \text{ and } B) + P(A \text{ and } \bar{B}) + P(\bar{A} \text{ and } B). \qquad (3)$$

Adding equalities (1) and (2) memberwise and taking equality (3) into consideration, we easily find that

$$P(A) + P(B) = P(A \text{ and } B) + P(A \text{ or } B),$$

from which it follows that

$$P(A \text{ or } B) = P(A) + P(B) - P(A \text{ and } B). \qquad (4)$$

We have arrived at a very important result. Although we carried out our reasoning for a particular example, it was so general that the result can be considered established for any pair of events A and B. Up to this point, we obtained expressions for probabilities $P(A \text{ or } B)$ only under very particular assumptions concerning the connection between the events A and B (we first assumed them to be incompatible and, later, to be mutually independent). Formula (4) which we just obtained holds without any additional assumptions for an arbitrary pair of events A and B. It is true that we must not forget one essential difference between formula (4) and our previous formulas. In previous formulas, the probability $P(A \text{ or } B)$ was always expressed in terms of the probabilities $P(A)$ and $P(B)$, so that, knowing only the probabilities of the events A and B, we were always able to determine the probability of the event $(A \text{ or } B)$ uniquely. The situation is different in formula (4): to compute the quantity $P(A \text{ or } B)$ by this formula it is necessary to know, besides $P(A)$ and $P(B)$, the probability $P(A \text{ and } B)$, i.e., the probability of the simultaneous occurrence of the events A and B. To find this same probability in the general case, with arbitrary connection between the events A and B, is usually no easier than to find $P(A \text{ or } B)$; therefore, for practical calculations we seldom use formula (4) directly—but it is, nonetheless, of very great theoretical significance.

We shall first convince ourselves that our previous formulas can easily be obtained from formula (4) as special cases. If the events A and B are mutually incompatible, then the event $(A \text{ and } B)$ is impossible—hence, $P(A \text{ and } B) = 0$—and formula (4) leads to the relation

$$P(A \text{ or } B) = P(A) + P(B),$$

i.e., to the addition law. If the events A and B are mutually independent, then, according to formula (3) on page 22, we have

$$P(A \text{ and } B) = P(A) \cdot P(B),$$

and formula (4) yields

$$P(A \text{ or } B) = P(A) + P(B) - P(A) \cdot P(B)$$
$$= 1 - [1 - P(A)] \cdot [1 - P(B)].$$

Thus, we obtain formula (5) on page 25 (for the case $n = 2$).

Furthermore, we deduce an important corollary from formula (4). Since $P(A \text{ and } B) \geq 0$ in all cases, it follows from formula (4) in all cases that

$$P(A \text{ or } B) \leq P(A) + P(B). \tag{5}$$

This inequality can easily be generalized to any number of events. Thus, for instance, in the case of three events, we have, by virtue of (5),

$$P(A \text{ or } B \text{ or } C) \leq P(A \text{ or } B) + P(C)$$
$$\leq P(A) + P(B) + P(C),$$

and, clearly, one can proceed in the same way from three events to four, and so on.   We obtain the following general result:

The probability of the occurrence of at least one of several events never exceeds the sum of the probabilities of these events.

In this connection, the equality sign holds only in the case when every pair of the given events is mutually incompatible.

§ 11. Formula for total probability

We return once more to the bulb example on page 18 and use, for the various results of the experiments, the notation introduced on page 27.   The probability that a bulb is of standard quality under the condition that it was manufactured at the second plant equals, as we have already seen more than once,

$$P_{\bar{B}}(A) = \frac{189}{300} = 0.63$$

and the probability of the same event under the condition that the bulb was manufactured at the first plant is

$$P_B(A) = \frac{581}{700} = 0.83.$$

Let us assume that these two numbers are known and that we also know that the probability that the bulb was manufactured at the first plant is

$$P(B) = 0.7$$

and at the second plant is

$$P(\bar{B}) = 0.3.$$

It is required that one find the unconditional probability $P(A)$, i.e., the probability that a random bulb is of standard quality, without any assumptions concerning the place where it was manufactured.

In order to solve this problem, we shall reason as follows.   We denote by E the joint event consisting of 1) that the bulb was issued by the first plant and 2) that it is standard, and by F the analogous event for the second plant.   Since every standard bulb is manufactured by the first or second plant, the event A is equivalent to the event "E or F" and since the events E and F are mutually incompatible, we have, by the addition law

$$P(A) = P(E) + P(F). \tag{6}$$

On the other hand, in order that the event E hold, it is necessary 1) that the bulb be manufactured by the first plant (B) and 2) that it be standard (A); therefore, the event E is equivalent to the event "B and A," from which it follows, by the multiplication rule, that

$$P(E) = P(B) \cdot P_B(A).$$

In exactly the same way we find that

$$P(F) = P(\overline{B}) \cdot P_{\overline{B}}(A),$$

and, substituting these expressions into equality (6), we have

$$P(A) = P(B) \cdot P_B(A) + P(\overline{B}) \cdot P_{\overline{B}}(A).$$

This formula solves the problem we posed. Substituting the given numbers, we find that $P(A) = 0.77$.

EXAMPLE. For a seeding, there are prepared wheat seeds of the variety I containing as admixture small quantities of other varieties—II, III, IV. We take one of these grains. The event that this grain is of variety I will be denoted by A_1, that it is of variety II by A_2, of variety III by A_3, and, finally, of variety IV by A_4. It is known that the probability that a grain taken at random turns out to be of a certain variety equals:

$$P(A_1) = 0.96; \quad P(A_2) = 0.01; \quad P(A_3) = 0.02 \quad P(A_4) = 0.01.$$

(The sum of these four numbers equals unity, as it should in every case of a complete system of events.)

The probability that a spike containing no less than 50 grains will grow from the grain equals:

1) 0.50 for a grain of variety I
2) 0.15   ,,   ,,   II
3) 0.20   ,,   ,,   III
4) 0.05   ,,   ,,   IV.

It is required that one find the unconditional probability that the spike has no less than 50 grains.

Let K be the event that the spike contains no less than 50 grains; then, by the condition of the problem, we have

$$P_{A_1}(K) = 0.50; \quad P_{A_2}(K) = 0.15; \quad P_{A_3}(K) = 0.20; \quad P_{A_4}(K) = 0.05.$$

Our problem is to determine $P(K)$. We denote by E_1 the event that the grain turns out to be of variety I and that the spike growing from it will contain no less than 50 grains, so that E_1 is equivalent to the event $(A_1$ and $K)$; in the same way, we denote

the event $(A_2$ and $K)$ by E_2

the event $(A_3$ and $K)$ by E_3

the event $(A_4$ and $K)$ by E_4.

Obviously, for the event K to occur it is necessary that one of the events E_1, E_2, E_3, or E_4 occur and since any pair of these events is mutually incompatible, we obtain, by the addition rule

$$P(K) = P(E_1) + P(E_2) + P(E_3) + P(E_4). \qquad (7)$$

On the other hand, according to the multiplication rule, we have

$$P(E_1) = P(A_1 \text{ and } K) = P(A_1) \cdot P_{A_1}(K)$$
$$P(E_2) = P(A_2 \text{ and } K) = P(A_2) \cdot P_{A_2}(K)$$
$$P(E_3) = P(A_3 \text{ and } K) = P(A_3) \cdot P_{A_3}(K)$$
$$P(E_4) = P(A_4 \text{ and } K) = P(A_4) \cdot P_{A_4}(K).$$

Substituting these expressions into formula (7), we find that

$$P(K) = P(A_1) \cdot P_{A_1}(K) + P(A_2) \cdot P_{A_2}(K)$$
$$+ P(A_3) \cdot P_{A_3}(K) + P(A_4) \cdot P_{A_4}(K),$$

which obviously solves our problem. Substituting the given numbers into the last equation, we find that

$$P(K) = 0.486.$$

The two examples which we considered here in detail bring us to an important general rule which we can now formulate and prove without difficulty. Suppose a given operation admits of the results $A_1, A_2, \ldots, A_n$ and that these form a complete system of events. (Let us recall that this means that any two of these events are mutually incompatible and that some one of them must necessarily occur.) Then for an arbitrary possible result K of this operation, the relation

$$P(K) = P(A_1) \cdot P_{A_1}(K) + P(A_2) \cdot P_{A_2}(K) + \ldots + P(A_n) \cdot P_{A_n}(K) \quad (8)$$

holds. Rule (8) is usually called the "*formula for total probability.*" Its proof is carried out exactly as in the two examples we considered above: first, the occurrence of the event K requires the occurrence of one of the events "A_i and K" so that, by the addition rule, we have

$$P(K) = \sum_{i=1}^{n} P(A_i \text{ and } K); \qquad (9)$$

second, by the multiplication rule,

$$P(A_i \text{ and } K) = P(A_i) \cdot P_{A_i}(K);$$

substituting these expressions into equation (9) we arrive at formula (8).

§ 12. Bayes's formula

The formulas of the preceding section enable us to derive an important result having numerous applications. We start with a formal derivation, postponing an explanation of the real meaning of the final formula until we consider examples.

Again, let the events $A_1, A_2, \ldots, A_n$ form a complete system of results of some operation. Then, if K denotes an arbitrary result of this operation, we have, by the multiplication rule

$$P(A_i \text{ and } K) = P(A_i) \cdot P_{A_i}(K) = P(K) \cdot P_K(A_i) \quad (1 \le i \le n),$$

from which it follows that

$$P_K(A_i) = \frac{P(A_i) \cdot P_{A_i}(K)}{P(K)} \quad (1 \le i \le n),$$

or, expressing the denominator of the fraction obtained according to the formula for total probability (8) in the preceding section, we find that

$$P_K(A_i) = \frac{P(A_i) \cdot P_{A_i}(K)}{\sum\limits_{r=1}^{n} P(A_r) \cdot P_{A_r}(K)} \quad (1 \le i \le n). \qquad (10)$$

This is *Bayes's formula*, which has many applications in practice in the calculation of probabilities. We apply it most frequently in situations illustrated by the following example.

Suppose a target situated on a linear segment MN (see Fig. 3) is being fired upon; we imagine the segment MN to be subdivided into five small subsegments a, b', b'', c', c''. We assume that the precise position of the target is not known; we only know the probability that the target lies on one or another of these subsegments. We suppose these probabilities are equal to

$$P(a) = 0.48; \quad P(b') = P(b'') = 0.21; \quad P(c') = P(c'') = 0.05,$$

where now a, b', b'', c', c'' denote the following events: the target lies in the segment a, b', b'', c', c'', respectively. (Note that the sum of these numbers equals unity.) The largest probability corresponds to the segment a toward which we therefore, naturally, aim our shot.

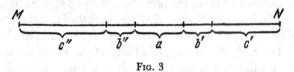

Fig. 3

However, due to unavoidable errors in firing, the target can also be destroyed when it is not in a but in any of the other segments. Suppose the probability of destroying the target (event K) is

$$P_a(K) = 0.56 \text{ if the target lies in the segment } a$$
$$P_{b'}(K) = 0.18 \quad ,, \quad ,, \quad ,, \quad ,, \quad b'$$
$$P_{b''}(K) = 0.16 \quad ,, \quad ,, \quad ,, \quad ,, \quad b''$$
$$P_{c'}(K) = 0.06 \quad ,, \quad ,, \quad ,, \quad ,, \quad c'$$
$$P_{c''}(K) = 0.02 \quad ,, \quad ,, \quad ,, \quad ,, \quad c''.$$

We assume that a shot has been fired and that the target was destroyed (i.e., event K occurred). As a result of this, the probabilities of the various positions of the target which we had earlier [i.e., the numbers $P(a)$, $P(b')$, ...] must be recalculated. The qualitative aspect of this revised calculation is clear without any computations, for we shot at the segment a and hit the target—it is clear that the probability $P(a)$ in this connection must increase. Now we wish to compute exactly and quantitatively the new value due to our shot; i.e., we wish to find an exact expression for the probabilities $P_K(a)$, $P_K(b')$, ... of the various possible positions of the target under the condition that the target was destroyed by the shot fired. Bayes's formula (10) at once gives us the answer to this problem. Thus,

$$P_K(a) = \{P(a) \cdot P_a(K)\}/\{P(a) \cdot P_a(K) + P(b') \cdot P_{b'}(K)$$
$$+ P(b'') \cdot P_{b''}(K) + P(c') \cdot P_{c'}(K) + P(c'') \cdot P_{c''}(K)\} \approx 0.8;$$

we see that $P_K(a)$ is in fact larger than $P(a)$.

We easily find the probabilities $P_K(b')$, ... for the other positions of the target in a similar manner. For the calculations, it is useful to note that the expressions given for these probabilities by Bayes's formula differ from one another in their numerators while the denominators in these expressions are, however, the same, $P(K) \approx 0.34$.

The general scheme of this type of situation can be described as follows. The conditions of the operation contain some unknown element with respect to which n distinct "hypotheses" can be made: $A_1, A_2, \ldots, A_n$ which form a complete system of events. For one reason or another we know the probabilities $P(A_i)$ of these hypotheses to be tested; it is also known that the hypothesis A_i "conveys" a probability $P_{A_i}(K)$ $(1 \leq i \leq n)$ to some event K (for instance, hitting a target). Here, $P_{A_i}(K)$ is the probability of the event K calculated under the condition that the hypothesis A_i is true. If, as the result of a trial, event K has occurred, then this requires a re-evaluation of the probability of the hypothesis A_i and the problem consists in finding the new probabilities $P_K(A_i)$ of these hypotheses; Bayes's formula gives the answer.

In artillery practice, so-called test-firings are carried out which have for their purpose making more precise our knowledge of the firing conditions. In this regard, not only the position of the target can serve as the unknown element whose effect is required to be made precise, but also any other element in the firing conditions which influences the effectiveness of the results (in particular, some peculiarity of the fire-arm used). It often happens that not one such shot is fired but, rather, several, and the problem posed is to calculate the new probabilities of the hypotheses on the basis of the firing results obtained. In all such cases, Bayes's formula also easily solves the problems.

For the sake of brevity in writing, we shall set, in the general scheme considered by us,

$$P(A_i) = P_i \text{ and } P_{A_i}(K) = p_i \quad (1 \leq i \leq n),$$

so that Bayes's formula has the simple form

$$P_K(A_i) = \frac{P_i p_i}{\sum\limits_{r=1}^{n} P_r p_r}.$$

We assume that s test shots have been fired, in which connection the result K occurred m times and did not occur $s-m$ times. We denote by K^* the result obtained from a series of s shots. We can assume that the results of individual shots constitute mutually independent events. If the hypothesis A_i is valid, the probability of the result K equals p_i and, hence, the probability of the complementary event that K does not occur equals $1-p_i$.

The probability that the result K occurred for the definite m shots equals $p_i^m(1-p_i)^{s-m}$ according to the multiplication rule for independent events. Since the m shots in which the result K occurred can be any of the s fired, the event K^* can be realized in C_s^m incompatible ways. Thus, according to the rule for the addition of probabilities, we have

$$P_{A_i}(K^*) = C_s^m p_i^m (1-p_i)^{s-m} \quad (1 \leq i \leq n),$$

and Bayes's formula yields

$$P_{K^*}(A_i) = \frac{P_i p_i^m (1-p_i)^{s-m}}{\sum\limits_{r=1}^{n} P_r p_r^m (1-p_r)^{s-m}} \quad (1 \leq i \leq n), \qquad (11)$$

which solves the problem posed. Of course, such problems arise not only in artillery practice, but also in other areas of human activity.

EXAMPLE 1.   Referring to the problem we considered in the beginning of the present section, we now seek the probability that the target lies in the segment a if two successive shots at this segment yielded hits.

Denoting by K^* the event of hitting the target twice, we have, according to formula (11)

$$P_{K^*}(a) = \frac{P(a) \cdot [P_a(K)]^2}{P(a) \cdot [P_a(K)]^2 + P(b') \cdot [P_{b'}(K)]^2 + \ldots}.$$

We leave it to the reader to carry out the uncomplicated calculation and verify that as a result of hitting the target twice the probability that the target is situated in the segment a has been increased still more.

EXAMPLE 2.   The probability that in a certain production process the articles satisfy a prescribed standard equals 0.96. A simplified system of testing[1] is suggested which for the articles satisfying the standard yield a positive result with probability 0.98 and for articles which do not satisfy the standard a positive result with a probability 0.05. What is the probability that the articles which endure the simplified test twice satisfy the standard?

Here, a complete system of hypotheses consists of two complementary events: 1) that the article satisfies the standard, or 2) that the article does not satisfy the standard. The probabilities of these hypotheses are, before the test, equal to $P_1 = 0.96$ and $P_2 = 0.04$,

---

[1] The necessity for a simplified control is encountered very frequently in practice. For instance, if upon dispensing electric light bulbs all of them were subjected to testing for their ability to burn for a period, say, of not less than 1200 hours, then the consumer would obtain only burnt-out or almost burnt-out bulbs. Thus one must replace the test for period of burning by other tests—for example, testing the bulb for lighting up.

respectively. Under the first hypothesis, the probability that the article endures the test equals $p_1 = 0.98$ and, under the second hypothesis, the probability equals $p_2 = 0.05$. After a two-fold test, the probability of the first hypothesis is equal, on the basis of formula (11), to

$$\frac{P_1 p_1^2}{P_1 p_1^2 + P_2 p_2^2} = \frac{0.96 \cdot (0.98)^2}{0.96 \cdot (0.98)^2 + 0.04 \cdot (0.05)^2} \approx 0.9999.$$

We see that if the article endured the test indicated in the conditions of the problem, then we can make an error only once in ten thousand cases assuming that it is standard. This, of course, completely satisfies the requirements in practice.

EXAMPLE 3. In an examination of a patient, it is suspected that he has one of three illnesses: A_1, A_2, A_3. Their probabilities, under prescribed conditions, are

$$P_1 = 1/2, \quad P_2 = 1/6, \quad P_3 = 1/3,$$

respectively. In order to make the diagnosis more precise, some analysis is specified which yields a positive result with probability 0.1 in the case of illness A_1, with probability 0.2 in the case of illness A_2, and with probability 0.9 in the case of illness A_3. The analysis was carried out five times and yielded a positive result four times and a negative result once. It is required that one find the probability of each of the illnesses after the analysis.

In the case of illness A_1, the probability of the indicated results of the analyses is equal, by the multiplication rule, to $p_1 = C_5^4 (0.1)^4 \cdot 0.9$. For the second hypothesis, this probability equals $p_2 = C_5^4 (0.2)^4 \cdot 0.8$ and for the third it is equal to $p_3 = C_5^4 (0.9)^4 \cdot 0.1$.

According to Bayes's formula, we find that after the analyses the probability of illness A_1 turns out to be equal to

$$\frac{P_1 p_1}{P_1 p_1 + P_2 p_2 + P_3 p_3}$$

$$= \frac{(1/2) \cdot (0.1)^4 \cdot 0.9}{(1/2) \cdot (0.1)^4 \cdot 0.9 + (1/6) \cdot (0.2)^4 \cdot 0.8 + (1/3) \cdot (0.9)^4 \cdot 0.1} \approx 0.002;$$

the probability of illness A_2 is

$$\frac{P_2 p_2}{P_1 p_1 + P_2 p_2 + P_3 p_3}$$

$$= \frac{(1/6) \cdot (0.2)^4 \cdot 0.8}{(1/2) \cdot (0.1)^4 \cdot 0.9 + (1/6) \cdot (0.2)^4 \cdot 0.8 + (1/3) \cdot (0.9)^4 \cdot 0.1} \approx 0.01;$$

and for illness A_3 it is

$$\frac{P_3p_3}{P_1p_1+P_2p_2+P_3p_3}$$

$$= \frac{(1/3)\cdot(0.9)^4\cdot0.1}{(1/2)\cdot(0.1)^4\cdot0.9+(1/6)\cdot(0.2)^4\cdot0.8+(1/3)\cdot(0.9)^4\cdot0.1} \approx 0.988.$$

Since these three events A_1, A_2, A_3 form, even after the test, a complete system of events, we can as a check on the calculation carried out add the three numbers obtained and verify that their sum is equal to unity, as before.

CHAPTER 5

BERNOULLI'S SCHEME

§ 13. Examples

EXAMPLE 1. Among fibers of cotton of a definite sort 75% on the average have lengths less than 45 mm. and 25% have lengths greater than (or equal to) 45 mm. Find the probability that of three fibers taken at random two will be shorter than and one will be longer than 45 mm.

We denote the event of choosing a fiber of length less than 45 mm. by A and the event of choosing a fiber of length greater than 45 mm. by B; it is then clear that

$$P(A) = 3/4; \quad P(B) = 1/4.$$

We shall further agree to denote the following compound event by AAB: the first two fibers chosen are shorter than 45 mm. and the third fiber is longer than 45 mm. It is clear what the meaning of the schemes BBA, ABA, and so on, will be. Our problem is to compute the probability of the event C: that of three fibers two are shorter than 45 mm. and one fiber is longer than 45 mm. Evidently, for this to happen one of the following schemes must be realized:

$$AAB, ABA, BAA. \tag{1}$$

Since any two of these three results are mutually incompatible we have, by the addition rule

$$P(C) = P(AAB) + P(ABA) + P(BAA).$$

All three terms in the right member are equal inasmuch as the results of the choice of the fibers can be assumed to be mutually independent events. The probability of each of the schemes (1), according to the multiplication rule for probabilities of independent events, is representable as the product of three factors of which two equal $P(A) = 3/4$ and one equals $P(B) = 1/4$. Thus, the probability of each of the three schemes (1) equals

$$(3/4)^2 \cdot (1/4) = 9/64,$$

and, consequently,

$$P(C) = 3 \cdot (9/64) = 27/64,$$

which is the solution of our problem.

EXAMPLE 2. As the result of observations extending over many decades it was found that of every 1000 newly born children on the average there are born 515 boys and 485 girls. In a certain family there are six children. Find the probability that there are no more than two girls among them.

For the occurrence of the event whose probability we are seeking, it is necessary that there be either 0 or 1 or 2 girls. The probabilities of these particular events will be denoted by P_0, P_1, P_2, respectively. It is clear that, according to the rule for the addition of probabilities, the probability sought is

$$P = P_0 + P_1 + P_2. \tag{2}$$

For each child, the probability that it is a boy equals 0.515 and, hence, the probability that it is a girl equals 0.485.

P_0 is the easiest to find; this is the probability that all the children in the family are boys. Since the birth of a child of either sex can be considered as independent of the sex of the remaining children, the probability, according to the rule for the multiplication of probabilities, that all six children are boys is equal to the product of six factors each equal to 0.515, i.e.,

$$P_0 = (0.515)^6 \approx 0.018.$$

We now go over to the calculation of P_1, i.e., the probability that of the six children in the family one child is a girl and the remaining five are boys. This event can occur in six different ways depending on which child in the order of birth is a girl (i.e., first, second, etc.). We consider any of the possible ways of this event, for example the one that a girl is born as the fourth child. The probability of this possibility, according to the multiplication rule, equals the product of six factors of which five equal 0.515 and the sixth (situated in the fourth place) equals 0.485; i.e., this probability equals $(0.515)^5 \cdot 0.485$. This is also the probability of each of the other five possibilities of the event which interests us at the moment; therefore, the probability P_1 of this event is equal, according to the addition rule, to the sum of six numbers each equal to $(0.515)^5 \cdot 0.485$, i.e.,

$$P_1 = 6 \cdot (0.515)^5 \cdot 0.485 \approx 0.105.$$

We now turn to the calculation of P_2 (i.e., the probability that two of the children are girls and four are boys). Analogous to what precedes, we at once note that this event admits of a whole series of possibilities. One of the possibilities will be, for instance, the following: the second and fifth child in order of birth are girls and the remainder are boys. The probability of each of the possibilities, according to the multiplication rule, equals $(0.515)^4 \cdot (0.485)^2$ and, consequently, P_2 equals, by the addition rule, the number $(0.515)^4 \cdot (0.485)^2$, multiplied by the number of all possibilities of the type considered; the entire problem thus reduces to the determination of this last number.

Each of the possibilities is characterized by the fact that of six children two are girls and the remainder are boys; the number of different possibilities consequently equals the number of distinct choices of two children from the six at hand. The number of such choices equals the number of combinations of six distinct objects taken two at a time; i.e., $C_6^2 = (6 \cdot 5)/(2 \cdot 1) = 15$. Thus,

$$P_2 = C_6^2 \cdot (0.515)^4 \cdot (0.485)^2 = 15 \cdot (0.515)^4 \cdot (0.485)^2 \approx 0.247.$$

Combining the results obtained above, we have

$$P = P_0 + P_1 + P_2 \approx 0.018 + 0.105 + 0.247 = 0.370.$$

Thus, in about 37% of the families having six children we will find fewer than three girls and, hence, more than three boys among the children.

§ 14. The Bernoulli formulas

In the preceding section, we became acquainted by means of a number of examples with the scheme of *repeated trials*, in each of which an event A can be realized. We attribute a very broad and variegated sense to the word "trial." Thus, if we fire at a certain target, by a trial we shall understand each individual shot. If we test electric light bulbs for length of burning, then a trial will be understood to be the testing of each bulb. If we are studying the composition of newly born children by sex, weight, or height, then a trial will be understood to be the investigation of an individual child. In general, by a trial we shall in what follows understand the realization of certain conditions in the presence of which some event of interest to us can occur.

We have now arrived at the consideration of one of the important schemes in the theory of probability having, besides application in various branches of knowledge, great significance also in probability theory itself as a mathematical science. This scheme consists in considering a sequence of mutually independent trials, i.e., of such trials for which the probability of some result in each of them does not depend on what results occurred or will occur in the remainder.   In each of these trials, there can occur (or not occur) some event A with probability p which does not depend on the number of trials.   The scheme just described has received the name *Bernoulli scheme* since the origin of its systematic study can be traced back to the renowned Swiss mathematician Jacob Bernoulli, who lived at the end of the seventeenth century.

We have already dealt with the Bernoulli scheme in our examples; in order to convince ourselves of this, it is sufficient to recall the examples of the preceding section.   We shall now solve the following general problem; all the examples we considered up to this point in this chapter were particular cases of this.

PROBLEM.   Under certain conditions, the probability that the event A occurs in every trial equals p; find the probability that a sequence of n independent trials yields k occurrences and $n-k$ nonoccurrences of the event A.

The event whose probability is sought splits into a number of possibilities; in order to obtain one definite possibility, we must arbitrarily choose from the given sequence any k trials and assume that the event A occurred for precisely these k trials and that A did not occur for the remaining $n-k$.   Thus, every such possibility requires the occurrence of n definite results—in this number k occurrences and $n-k$ nonoccurrences of the event A.   By the multiplication rule, we find that the probability of each definite possibility equals

$$p^k(1-p)^{n-k}.$$

The number of different possibilities equals the number of different sets of k trials each of which can be constructed from n distinct trials, i.e., it is equal to C_n^k.   Applying the addition rule and the known formula for the number of combinations of n objects taken k at a time,

$$C_n^k = \frac{n(n-1)\ldots[n-(k-1)]}{k(k-1)\ldots2\cdot1},$$

we find that the probability of k occurrences of the event A with n independent trials equals

$$P_n(k) = \frac{n(n-1)\ldots[n-(k-1)]}{k(k-1)\ldots2\cdot1}\,p^k(1-p)^{n-k}, \tag{3}$$

which solves our problem.   It is frequently more convenient to represent the expression C_n^k in a somewhat different form; multiplying the numerator and denominator by the product $(n-k)[n-(k+1)]$ $\ldots 2 \cdot 1$, we obtain

$$C_n^k = \frac{n(n-1)\ldots 2 \cdot 1}{k(k-1)\ldots 2 \cdot 1(n-k)\cdot[n-(k+1)]\ldots 2 \cdot 1},$$

or, denoting for brevity the product of all integers from 1 to m inclusively by $m!$,

$$C_n^k = \frac{n!}{k!(n-k)!}.$$

For $P_n(k)$, this yields

$$P_n(k) = \frac{n!}{k!(n-k)!}\, p^k(1-p)^{n-k}. \tag{4}$$

Formulas (3) and (4) are usually called *Bernoulli's formulas*. For large values of n and k, the computation of $P_n(k)$ according to these formulas is rather difficult since the factorials $n!$, $k!$, $(n-k)!$ are very large numbers which are rather cumbersome to evaluate. Therefore, in calculations of this type specially compiled tables of factorials as well as various approximation formulas are extensively used.

EXAMPLE. The probability that the consumption of water at a certain factory is normal (i.e., it is not more than a prescribed number of liters every twenty-four hours) equals 3/4. Find the probability that in the next 6 days the consumption of water will be normal in the course of 0, 1, 2, 3, 4, 5, 6 days.

Denoting by $P_6(k)$ the probability that in the course of k days out of 6 the consumption of water will be normal, we find, by formula (3) (where we must set $p=3/4$), that

$$P_6(6) = (3/4)^6 = 3^6/4^6,$$

$$P_6(5) = 6 \cdot (3/4)^5 \cdot 1/4 = \frac{6 \cdot 3^5}{4^6},$$

$$P_6(4) = C_6^4 \cdot (3/4)^4 \cdot (1/4)^2 = C_6^2 \frac{3^4}{4^6} = \frac{6 \cdot 5}{2 \cdot 1} \cdot \frac{3^4}{4^6} = \frac{15 \cdot 3^4}{4^6},$$

$$P_6(3) = C_6^3 \cdot (3/4)^3 \cdot (1/4)^3 = \frac{6 \cdot 5 \cdot 4}{3 \cdot 2 \cdot 1} \cdot \frac{3^3}{4^6} = \frac{20 \cdot 3^3}{4^6},$$

$$P_6(2) = \frac{6 \cdot 5}{2 \cdot 1}\,(3/4)^2 \cdot (1/4)^4 = \frac{15 \cdot 3^2}{4^6},$$

$$P_6(1) = 6 \cdot (3/4) \cdot (1/4)^5 = \frac{6 \cdot 3}{4^6};$$

finally, we evidently have $P_6(0)$ (i.e., the probability that there is excessive consumption in each of the 6 days) equal to $1/4^6$. All six probabilities are expressed as fractions with the same denominator, $4^6 = 4096$; we use this, of course, to shorten our calculations.   These yield

$$P_6(6) \approx 0.18; \quad P_6(5) \approx 0.36; \quad P_6(4) = 0.30;$$
$$P_6(3) \approx 0.13; \quad P_6(2) \approx 0.03; \quad P_6(1) = P_6(0) \approx 0.00.$$

We see that it is most probable that there will be an excessive consumption of water in the course of one or two days of the six and that the probability of excessive consumption in the course of five or six days, i.e., $P_6(1) + P_6(0)$, practically equals zero.

§ 15. The most probable number of occurrences of an event

The example which we just considered shows that the probability of a normal consumption of water in the course of exactly k days with increasing k at first increases and then, having attained its largest value, begins to decrease; this is most clearly seen if the variation of the probability $P_6(k)$ with increasing k is expressed geometrically in the form of a diagram, shown in Fig. 4.   A still clearer picture is

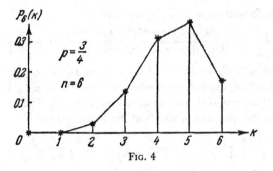

FIG. 4

given by diagrams of the variation of the quantity $P_n(k)$ as k increases when the number n becomes larger; thus, for $n = 15$ and $p = 1/2$, the diagram has the form shown in Fig. 5.

In practice, it is sometimes required to know what number of occurrences of the event is *most probable*, i.e., for what number k the probability $P_n(k)$ is the largest.   (In this connection, it is, of course, assumed that p and n are prescribed.)   The Bernoulli formulas allow us in all cases to find a simple solution of this problem; we shall occupy ourselves with this now.

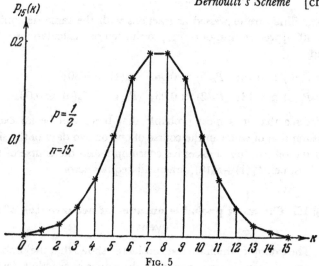

$P_{15}(\kappa)$

$p = \dfrac{1}{2}$

$n = 15$

FIG. 5

We first calculate the magnitude of the ratio $P_n(k+1)/P_n(k)$. By virtue of formula (4),

$$P_n(k+1) = \frac{n!}{(k+1)!(n-k-1)!}\, p^{k+1}(1-p)^{n-k-1}, \qquad (5)$$

and, from formulas (3) and (5), we have

$$\frac{P_n(k+1)}{P_n(k)} = \frac{n!\,k!\,(n-k)!\,p^{k+1}(1-p)^{n-k-1}}{(k+1)!\,(n-k-1)!\,n!\,p^{k}(1-p)^{n-k}} = \frac{n-k}{k+1}\cdot\frac{p}{1-p}.$$

The probability $P_n(k+1)$ will be larger than, equal to, or less than the probability $P_n(k)$ depending on whether or not the ratio $P_n(k+1)/P_n(k)$ is larger than, equal to, or less than unity, and the latter, as we see, reduces to the question of which of the three relations

$$\frac{n-k}{k+1}\cdot\frac{p}{1-p} > 1, \quad \frac{n-k}{k+1}\cdot\frac{p}{1-p} = 1, \quad \frac{n-k}{k+1}\cdot\frac{p}{1-p} < 1 \qquad (6)$$

is valid. If we wish, for example, to determine the values of k for which the inequality $P_n(k+1) > P_n(k)$ is satisfied, then we must recognize for what values of k the inequality

$$\frac{n-k}{k+1}\cdot\frac{p}{1-p} > 1$$

or

$$(n-k)p > (k+1)(1-p)$$

holds. From this we obtain

$$np - (1-p) > k;$$

thus, as long as k increases but does not attain the value $np-(1-p)$, we will always have $P_n(k+1) > P_n(k)$. Thus, with increasing k, the probability $P_n(k)$ will always increase. For example, in the scheme to which the diagram in Fig. 5 corresponds, we have $p=1/2$, $n=15$, $np-(1-p)=7$; this means that as long as $k<7$ (i.e., for all k from 0 to 6 inclusively), we have $P_n(k+1) > P_n(k)$. The diagram substantiates this.

In precisely the same way, starting with the other two relations in (6), we find that

$$P_n(k+1) = P_n(k) \text{ if } k = np-(1-p)$$

and

$$P_n(k+1) < P_n(k) \text{ if } k > np-(1-p);$$

thus, as soon as the number k exceeds the bound $np-(1-p)$, the probability $P_n(k)$ begins to decrease and will decrease to $P_n(n)$.

This derivation first of all convinces us that the behavior of the quantity $P_n(k)$ considered by us in the examples is a general law which holds in all cases: as the number k increases, $P_n(k)$ first increases and then decreases. But, more than this, this result also allows us to solve quickly the problem we have set for ourselves—i.e., to determine the most probable value of the number k. We denote this most probable value of the number k by k_0. Then

$$P_n(k_0+1) \leq P_n(k_0),$$

from which it follows, according to what precedes, that

$$k_0 \geq np-(1-p).$$

On the other hand,

$$P_n(k_0-1) \leq P_n(k_0),$$

from which, according to what precedes, the inequality

$$k_0-1 \leq np-(1-p)$$

or

$$k_0 \leq np-(1-p)+1 = np+p$$

must hold. Thus, the most probable value k_0 of the number k must satisfy the double inequality

$$np-(1-p) \leq k_0 \leq np+p. \tag{7}$$

The interval from $np-(1-p)$ to $np+p$, in which the number k_0 must therefore lie, has length 1 as can be shown by a simple calculation; therefore, if either of the endpoints of this interval, for instance the number $np-(1-p)$, is not an integer, then between these endpoints there will necessarily lie one, and only one, integer and k_0 will be uniquely determined. We ought to consider this case to be normal; for, p is less than 1, and therefore only in exceptional cases will the quantity $np - (1-p)$ be an integer. In this exceptional case, inequalities (7) yield two values for the number k_0: $np - (1-p)$ and $np + p$, which differ from one another by unity. Those two values will also be the most probable; their probabilities will be equal and exceed the probability of any other value of the number k. This exceptional case holds, for instance, in the scheme expressed by the diagram in Fig. 5; here, $n = 15$, $p = 1/2$ and hence $np - (1-p) = 7$, $np + p = 8$; the numbers 7 and 8 serve as the most probable values of the number k of occurrences of the event; their probabilities are equal to one another, each of them being approximately equal to 0.196. (All this can be seen on the diagram.)

EXAMPLE 1. As the result of observations over a period of many years, it was discovered, for a certain region, that the probability that rain falls on July 1 equals 4/17. Find the most probable number of rainy July 1's for the next 50 years. Here, $n = 50$, $p = 4/17$, and

$$np - (1-p) = 50 \cdot (4/17) - 13/17 = 11.$$

As this number turned out to be an integer, it means we are dealing with the exceptional case; the most probable value of the number of rainy days will be the numbers 11 and 12 which are equally probable.

EXAMPLE 2. In a physics experiment, particles of a prescribed type are being observed. Under fixed conditions, during an interval of time of definite length, on the average 60 particles appear and each of them has—with a probability 0.7—a velocity greater than v_0. Under other conditions, during the same interval of time there appear on the average only 50 particles, but for each of them the probability of having a velocity exceeding v_0 equals 0.8. Under what conditions of the experiment will the most probable number of particles having a velocity exceeding v_0 be the greatest?

Under the first conditions of the experiment,

$$n = 60, \quad p = 0.7, \quad np - (1-p) = 41.7, \quad k_0 = 42.$$

For the second conditions of the experiment,

$$n = 50, \quad p = 0.8, \quad np - (1-p) = 39.8, \quad k_0 = 40.$$

We see that the most probable number of "fast" particles under the first conditions of the experiment is somewhat larger than under the second.

In practice, we often encounter the situation when the number n is very large; e.g., in the case of mass firing, the mass production of articles, and so on. In this case the product np will also be a very large number provided the probability p is not unusually small. And since in the expressions $np - (1 - p)$ and $np + p$, between which lie the most probable number of occurrences of the event, the quantities p and $1 - p$ are less than unity, we see that both these expressions and hence the most probable number of occurrences of the event are all close to np. Thus, if the probability of completing a telephone connection in less than 15 seconds equals 0.74, then we can take $1000 \cdot 0.74$ as the most probable number of connections, among every 1000 calls coming into the central exchange, made in less than 15 seconds.

This result can be given a still more precise form. If k_0 denotes the most probable number of occurrences of the event in n trials, then k_0/n is the most probable "fraction" of occurrences of the event for the same n trials; inequalities (7) yield

$$p - \frac{1-p}{n} \leq \frac{k_0}{n} \leq p + \frac{p}{n}. \tag{8}$$

Let us assume that, leaving the probability p of the occurrence of the event for an individual trial invariant, we shall increase indefinitely the number of trials n. (In this connection we, of course, also increase the most probable number of occurrences k_0.) The fractions $(1 - p)/n$ and p/n, appearing in the left and right members of the inequalities (8) above will become smaller and smaller; this means that, for large n, these fractions can be disregarded. We can now consider both the left and right members of the inequalities (8) and hence also the fraction k_0/n contained between them to be equal to p. Thus, *the most probable ratio of occurrences of the event—provided there are a large number of trials—is practically equal to the probability of the occurrence of the event in an individual trial.*

For example, if for certain measurements the problem of making in an individual measurement an error comprised between α and β equals 0.84, then for a large number of measurements one can expect with the greatest probability errors comprised between α and β in approximately 84% of the cases. This does not mean, of course, that the probability of obtaining exactly 84% of such errors will be large; on the contrary, this "largest probability" itself will be very small in a large number of measurements (thus, we saw in the scheme in Fig. 5 that the largest probability turned out to be equal to 0.196 where we were dealing with 15 trials altogether; for a large number of trials it is significantly less). This probability is the largest only in the comparative sense: the probability of obtaining 84% of the measurements with errors comprised between α and β is larger than the probability of obtaining 83% or 86% of such measurements.

On the other hand, it is easily understandable that in extended series of measurements the probability of a certain individual number of errors of a given quantity cannot be of significant interest. For example, if we carry out 200 measurements, then it is doubtful whether it is expedient to calculate the probability that exactly 137 of them will be measurements with the prescribed precision because in practice it is immaterial whether the number is 137 or 136 or 138 or even, for instance, 140. In contrast, questions of the probability that the number of measurements for which the error is between prescribed bounds will be more than 100 of the 200 measurements made or that this number will be somewhere between 100 and 125 or that it will be less than 50, and so on, are certainly of practical interest. How should we express this type of probability? Suppose we wish, for example, to find the probability that the number of measurements will be between 100 and 120 (including 120); more specifically, we will seek the probability of satisfying the inequality

$$100 < k \leq 120,$$

where k is the number of measurements. For these inequalities to be realized, it is necessary that k be equal to one of the twenty numbers 101, 102, ..., 120. According to the addition rule, this probability equals

$$P(100 < k \leq 120) = P_{200}(101) + P_{200}(102) + \ldots + P_{200}(120).$$

To calculate this sum directly, we would have first to compute 20 individual probabilities of the type $P_n(k)$ according to formula (3); for such large numbers, such calculations present insurmountable difficulties. Therefore, sums of the form obtained are never computed by means of direct calculations in practice. For this purpose there exist suitable approximation formulas and tables. The composition of these formulas and tables is based on complicated methods of mathematical analysis, which we shall not touch upon here. However, concerning probabilities of the type $P(100 < k \leq 120)$ one can obtain information by simple lines of reasoning in many cases which lead to the complete solution of the problem posed. We shall discuss this problem in the following chapter.

CHAPTER 6

BERNOULLI'S THEOREM

§ 16. Content of Bernoulli's theorem

Let us take another good look at the diagram in Fig. 5 (on page 44), where the probabilities of various values of the number k of occurrences of the event under consideration are the numbers $P_{15}(k)$, which are depicted by the vertical lines. The probability assigned to some *segment* of values of k (the probability that the *number* of occurrences of the event of interest to us turns out to be equal to some one of the numbers of this segment) is equal, according to the addition rule, to the sum of the probabilities of all the numbers of this segment; i.e., it is equal to the sum of the lengths of all vertical lines situated over this segment. Pictorially, the figure shows that this sum is quite different for various segments of the same length. Thus, the segments $2 \leq k < 5$ and $7 \leq k < 10$ have the same length; the probability of each of them is expressed by the sum of the lengths of three vertical lines, and we see that for the second segment this sum is significantly larger than for the first. We already know that the diagrams of the probabilities $P_n(k)$ have, for all n, basically, the same form as the diagram in Fig. 5; i.e., the quantity $P_n(k)$ at first increases with increasing k and then, after passing through its largest value, it decreases. It is therefore clear that of the two segments of values of the number k having the same length, the one situated nearer the most probable value, k_0, will in all cases have the largest probability. In particular, on the segment having the number k_0 as its center we will always have a greater probability than on any other segment of the same length.

But it turns out that much more can be said in this regard. There are in all $n+1$ possible values of the number k of occurrences of the event in n trials $(0 \leq k \leq n)$. We take the segment having center at k_0 and containing only a small fractional part, for example one hundredth, of the possible values of the number k. It then turns out that if the total number n of trials is very large, the predominant probability will correspond to this segment and all other values of the number k taken together have a negligibly small probability. Thus, although the segment we chose is negligibly small in comparison with n (on the figure it occupies in all a one-hundredth part of the entire length of the diagram), nevertheless, the sum of the vertical lines situated over it will be significantly larger than the sum of all remaining vertical lines.   The reason for this lies in the fact that the lines in the central part of the diagram are many times larger than the lines situated near the ends.   Thus, for large n the diagram of the quantity $P_n(k)$ has a form which is approximately that shown in Fig. 6.

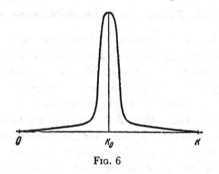

FIG. 6

In practice, this obviously means the following: *if we perform a series of a large number* n *of trials, then we can expect with a probability close to unity that the number* k *of occurrences of the event* A *will be very close to its most probable value, differing from the latter only by an insignificant fractional part of the total number* n *of trials made.*

This proposition, known under the name of *Bernoulli's theorem* and discovered at the beginning of the eighteenth century, is one of the important laws of probability theory.   Up to the middle of the last century, all proofs of this theorem required complicated mathematical means and the great Russian mathematician P. L. Chebyshev was the first to find a very simple and short derivation of this law; we now present Chebyshev's remarkable proof.

§ 17. Proof of Bernoulli's theorem

We already know that for a large number n of trials, the most probable number k_0 of occurrences of the event A differs very little from the quantity np, where p, as always, denotes the probability of the event A for an individual trial.   It is therefore sufficient for us to prove that, for a large number of trials, with very high probability the number k of occurrences of the event A will differ from np by very little —by not more than an arbitrarily small fractional part of the number n (not more, for example, than by 0.01 n or 0.001 n, or, in general, not more than by εn where ε is an arbitrarily small number).   In other words, we must show that the probability

$$P(|k-np| > \varepsilon n) \tag{1}$$

will be as small as we please for sufficiently large n.

In order to verify this, we note that according to the law of addition, probability (1) equals the sum of the probabilities $P_n(k)$ for all those

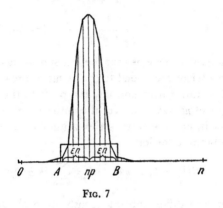

FIG. 7

values of the number k which lie at a distance not more than εn from np; in our typical diagram (Fig. 7), this sum is expressed by the sum of the lengths of all vertical lines lying exterior to the segment $\overline{AB}$—to the right as well as to the left of it.   Since the sum total of all the vertical lines (being the sum of the probabilities of a complete system of events) equals unity, this means that the overwhelming portion (almost equal to unity) of this sum corresponds to the segment $\overline{AB}$ and only a negligibly small part of it corresponds to the regions lying exterior to this segment.

Thus,

$$P(|k-np| > \varepsilon n) = \sum_{|k-np| > \varepsilon n} P_n(k). \tag{2}$$

We now turn to Chebyshev's line of reasoning.   Since in every term of the sum written down we have

$$\left| \frac{k-np}{\varepsilon n} \right| > 1$$

and hence

$$\left(\frac{k-np}{\varepsilon n} \right)^2 > 1,$$

we can only increase this sum if each of its terms $P_n(k)$ is replaced by
the expression

$$\left(\frac{k-np}{\varepsilon n}\right)^2 P_n(k).$$

Therefore,

$$P(|k-np| > \varepsilon n) < \sum_{|k-np| > \varepsilon n} \left(\frac{k-np}{\varepsilon n}\right)^2 P_n(k)$$

$$= \frac{1}{\varepsilon^2 n^2} \sum_{|k-np| > \varepsilon n} (k-np)^2 P_n(k).$$

Furthermore, it is obvious that the last sum is increased still more if
further new terms are added to the terms it already has, forcing the
number k to range over not only the parts to the left of $np-\varepsilon n$ and
to the right of $np+\varepsilon n$, but over the entire series of values which are
possible for it, i.e., the entire series of numbers from 0 to n inclusive.
We thus obtain, a fortiori,

$$P(|k-np| > \varepsilon n) < \frac{1}{\varepsilon^2 n^2} \sum_{k=0}^{n} (k-np)^2 P_n(k). \qquad (3)$$

The latter sum differs advantageously from all the preceding sums
in that it can be computed precisely; the Chebyshev method thus
consists of replacing sum (2), which is difficult to estimate, by the sum
(3), which admits of an exact computation.

We now proceed to make this calculation; no matter how long it
may appear to take us, these are simply difficulties of a technical
nature which anyone who knows algebra can handle. The remarkable idea of Chebyshev has already been completely utilized by us, as
it consisted, namely, in the transition from equality (2) to inequality (3).

First of all, we easily find that

$$\sum_{k=0}^{n} (k-np)^2 P_n(k) = \sum_{k=0}^{n} k^2 P_n(k) - 2np \sum_{k=0}^{n} k P_n(k) + n^2 p^2 \sum_{k=0}^{n} P_n(k). \quad (4)$$

Of the three sums in the right member, the last is equal to unity since
it is the sum of the probabilities of a complete system of events. This
means that it only remains for us to calculate the sums

$$\sum_{k=0}^{n} k P_n(k) \quad \text{and} \quad \sum_{k=0}^{n} k^2 P_n(k).$$

In this connection, in both sums the terms corresponding to $k=0$ are
equal to zero so that one can start the summation with $k=1$.

1) To calculate both sums, we express $P_n(k)$ according to formula (4), Chapter 5 (see page 42). We find that

$$\sum_{k=1}^{n} kP_n(k) = \sum_{k=1}^{n} \frac{kn!}{k!(n-k)!} p^k(1-p)^{n-k};$$

since, obviously, $n! = n(n-1)!$ and $k! = k(k-1)!$, we find that

$$\sum_{k=1}^{n} kP_n(k) = np \sum_{k=1}^{n} \frac{(n-1)!}{(k-1)![(n-1)-(k-1)]!} p^{k-1}(1-p)^{(n-1)-(k-1)},$$

or, setting $k-1=l$ in the sum in the right member and noting that l varies from 0 to $n-1$ as k varies from 1 to n,

$$\sum_{k=1}^{n} kP_n(k) = np \sum_{l=0}^{n-1} \frac{(n-1)!}{l!(n-1-l)!} p^l(1-p)^{n-1-l}$$

$$= np \sum_{l=0}^{n-1} P_{n-1}(l).$$

The last sum, i.e., $\sum_{l=0}^{n-1} P_{n-1}(l)$, of course, equals unity because it is the sum of the probabilities of a complete system of events—all possible numbers of occurrences of the event l for $n-1$ trials. Thus, for the sum $\sum_{k=0}^{n} kP(k)$, we obtain the very simple expression

$$\sum_{k=0}^{n} kP_n(k) = np. \tag{5}$$

2) To calculate the second sum, we first find the quantity $\sum_{k=1}^{n} k(k-1)P_n(k)$; since the term corresponding to $k=1$ is obviously equal to zero, the summation can begin with the value $k=2$. Noting that $n! = n(n-1)(n-2)!$ and that $k! = k(k-1)(k-2)!$, we easily conclude, setting $k-2=m$, similarly to what we did before, that

$$\sum_{k=1}^{n} k(k-1)P_n(k) = \sum_{k=2}^{n} k(k-1)P_n(k)$$

$$= \sum_{k=2}^{n} \frac{k(k-1)n!}{k!(n-k)!} p^k(1-p)^{n-k}$$

$$= n(n-1)p^2 \sum_{k=2}^{n} \frac{(n-2)!}{(k-2)![(n-2)-(k-2)]!} p^{k-2}$$

$$\cdot (1-p)^{(n-2)-(k-2)}$$

$$= n(n-1)p^2 \sum_{m=0}^{n-2} \frac{(n-2)!}{m!(n-2-m)!} p^m(1-p)^{n-2-m}$$

$$= n(n-1)p^2 \sum_{m=0}^{n-2} P_{n-2}(m) = n(n-1)p^2, \tag{6}$$

because the last sum is again equal to unity being the sum of the probabilities of a complete system of events—all possible numbers of occurrences of the events for $n-2$ trials.

Finally, formulas (5) and (6) yield

$$\sum_{k=1}^{n} k^2 P_n(k) = \sum_{k=1}^{n} k(k-1)P_n(k) + \sum_{k=1}^{n} kP_n(k)$$

$$= n(n-1)p^2 + np = n^2p^2 + np(1-p). \tag{7}$$

Now, both of the sums that we needed have been computed. Substituting results (5) and (7) into relation (4), we find finally that

$$\sum_{k=0}^{n} (k-np)^2 P_n(k) = n^2p^2 + np(1-p) - 2np \cdot np + n^2p^2$$

$$= np(1-p).$$

Substituting this simple expression we just derived into inequality (3), we obtain

$$P(|k-np| > \varepsilon n) < \frac{np(1-p)}{\varepsilon^2 n^2} = \frac{p(1-p)}{\varepsilon^2 n}. \tag{8}$$

This inequality completes the proof of everything required. In fact, it is true that we could have taken the number ε arbitrarily small; however, having chosen it, we do not change it any more. But the number n of trials in the sense of our assertion can be arbitrarily large. Therefore, the fraction $p(1-p)/(\varepsilon^2 n)$ can be assumed to be as small as we please, since with increasing n its denominator can be made arbitrarily large whereas the numerator at the same time remains unchanged.

For example, let $p=0.75$, so that

$$1-p = 0.25 \quad \text{and} \quad p(1-p) = 0.1875 < 0.2;$$

choose $\varepsilon = 0.01$; then inequality (8) yields

$$P\left(\left|k - \frac{3}{4}n\right| > 0.01n\right) < \frac{0.2}{0.0001 \cdot n} = \frac{2000}{n}.$$

If, for instance, we take $n=200,000$, then

$$P(|k - 150,000| > 2000) < 0.01.$$

In practice, this means, for example, the following: if in some production process, under fixed operating conditions, 75% on the average of the articles possess a certain property (for example, they belong to the first sort), then of 200,000 articles, from 148,000 to 152,000 articles will possess this property with a probability exceeding 0.99 (i.e., almost certainly).

In regard to this matter we must make two observations:

1. Inequality (8) yields a very rough estimate of the probability $P(|k-np| > \varepsilon n)$; in fact, this probability is significantly smaller—especially for large values of n. In practice, we therefore make use of more precise estimates whose derivation is, however, considerably more complicated.

2. The estimate, given by inequality (8), becomes significantly more precise when the probability p is very small—or just the opposite—very close to unity. Thus, if in the example we have just introduced, the probability that the article possesses a certain property equals $p = 0.95$, then $1-p = 0.05$, and $p(1-p) < 0.05$. Therefore, choosing $\varepsilon = 0.005$, $n = 200,000$, we find that

$$\frac{p(1-p)}{\varepsilon^2 n} < \frac{0.05 \cdot 1,000,000}{25 \cdot 200,000} = 0.01,$$

just as before. But now εn is not equal to 2000 but only to 1000; from this (since $np = 190,000$) we conclude that with practical certainty the number of articles possessing the property under consideration will, for a total number of 200,000 articles, lie between 189,000 and 191,000. Thus, inequality (8) practically guarantees us that the number of articles possessing the property concerned will be in an interval for $p = 0.95$ of half the length of that for $p = 0.75$, because we have here

$$P(|k-190,000| > 1000) < 0.01.$$

PROBLEM.   It is known that one-fourth of the workers in a particular branch of industry have an elementary school education. For a certain investigation, 200,000 workers are chosen at random. Find 1) the most probable value of the number of workers with an elementary school education among the 200,000 workers chosen and 2) the probability that the true (actual) number of such workers deviates from the most probable number by no more than 1.6%.

In the solution of this problem, we start with the fact that the probability of having an elementary education equals one-fourth for each of the 200,000 workers chosen at random. (This is precisely the key to the meaning of the phrase "at random.") Thus, in our problem, we have

$$n = 200,000, \quad p = 1/4, \quad k_0 = np = 50,000, \quad p(1-p) = 3/16.$$

We are seeking the probability that $|k-np| < 0.016np$ or that $|k-np| < 800$, where k is the number of workers with an elementary school education.   We choose ε so as to have $\varepsilon n = 800$; from this we find that $\varepsilon = 800/n = 0.004$.   Formula (8) yields

$$P(|k - 50{,}000| > 800) < \frac{3}{16 \cdot 0.000016 \cdot 200{,}000} \approx 0.06,$$

from which it follows that

$$P(|k - 50{,}000| < 800) > 0.94.$$

Answer.   The most probable value, which is what we are looking for, equals 50,000; the probability sought is greater than 0.94. (Actually, the probability sought is significantly closer to unity.)

PART II

RANDOM VARIABLES

CHAPTER 7

RANDOM VARIABLES AND DISTRIBUTION LAWS

§ 18. The concept of random variable

In our preceding discussion, we have many times now encountered quantities whose numerical values cannot be determined once for all but rather vary under the influence of random actions. Thus, the number of boys per hundred newly born babies will not be the same for every hundred. Or, the length of a cotton fiber of a definite sort varies significantly not only with the various regions where this sort is produced but even with the bush or boll itself. We now introduce still more examples of quantities of this kind.

1) Firing from the same firearm at the same target and under identical conditions, we nevertheless observe that the shells fall in different spots; this phenomenon is called the "dispersion" of shells. The distance of the spot where the shell falls from the place of its issue is a quantity which assumes various numerical values for various shots, depending on *random* conditions which could not have been taken into consideration in advance.

2) The speed of gas molecules does not remain constant, but rather varies due to collisions with other molecules. In view of the fact that every molecule can either collide or not collide with every other gas molecule, the variation of its speed possesses a purely *random* character.

3) The number of meteors falling onto the earth [these meteors are then called meteorites] in the course of a year, which enter the atmosphere and are not burned up, is not constant but rather is subject to significant variations which depend on a whole series of conditions of *random* character.

4) The weight of grains of wheat grown on a certain plot of ground is not equal to some definite quantity but varies from one grain to another. Because of the impossibility of evaluating the influence of *all* factors (e.g., the quality of the soil in the plot of ground on which the spike with the given grain grew, the conditions of sunlight under

which the grain was illuminated, the control of water, and others) determining the growth of the grain, its weight is a quantity which varies according to the case at hand.

Despite the diversity of the examples considered, all of them, from the point of view of interest to us, present the same picture. In each of these examples, we are dealing with a quantity which in one way or another characterizes the result of the operation undertaken (for example, the counting of meteors or the measurement of the length of fibers). Each of these quantities can assume, no matter how homogeneous we may strive to make their conditions, various values for various operations, which depend on random differences in the circumstances of these operations which are beyond our control. In probability theory, quantities of this sort are called *random* (or *stochastic*) *variables*; the examples we have introduced are already sufficient to convince us how important the study of random variables can be in the application of probability theory to the most varied areas of knowledge and practice.

To know a given random variable does not mean, of course, that we know its numerical value, because if we know, for instance, that a shell fell at a distance of 926 m. from the spot where it was fired, then by the same token this distance would already assume a definite numerical value and would cease to be a random variable. Then, what ought we to know about a random variable in order to have the most complete information concerning it, namely, as a *random* variable? Clearly, to this end, we must first of all know all the numerical values which it is capable of assuming. Thus, if in firing from a cannon under certain definite conditions the smallest range of the shell observed equals 904 m. and the greatest is 982 m., then the distance from the spot where the shell hits to the place where it is fired is capable of assuming all values included between these two bounds. In example 3), it is clear that the number of meteorites which reach the earth's surface in the course of a year can have as a value any non-negative integer, i.e., 0, 1, 2, and so forth.

However, knowledge of only one enumeration of possible values of a random variable still does not yield information about it which could serve as material for the estimates required in practice. Thus, if in the second example we consider the gas under two distinct temperatures, then the possible numerical values of the speed of the molecules for them are the same and, consequently, the enumeration of these values does not enable one to make any comparative estimate of these temperatures. Nevertheless, different temperatures indicate a very essential difference in the composition of the gas—a difference concerning which only one enumeration of the possible values of the speeds of the molecules does not give us any idea. If we wish to estimate the temperature of a given mass of gas and we are given only a list of possible values of the speeds of its molecules, then we naturally ask how often a certain speed is observed. In other words, we naturally strive to determine the *probabilities* of various possible values of the random variable of interest to us.

§ 19. The concept of law of distribution

As a beginning, we take a very simple example. A target, depicted in Fig. 8, is fired at; hitting region I gives the marksman three points, the region II two points, and the region III one point.[1]

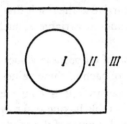

FIG. 8

As an example of a random variable, we consider the number of points won with a single shot. The numbers 1, 2, and 3 serve here as possible values; we denote the probabilities of these three values by p_1, p_2, p_3, respectively, so that, for instance, p_3 denotes the probability of hitting the region I of the target. Although the possible values of our random variable are the same for all marksmen, the probabilities p_1, p_2, and p_3 can differ very essentially from one another for different marksmen and, clearly, the difference in the quality of firing is determined by this difference. Thus, for a very good marksman, we could have, for instance, $p_3=0.8$, $p_2=0.2$, $p_1=0$; for an average marksman, $p_3=0.3$, $p_2=0.5$, $p_1=0.2$; and for a thoroughly unskilled marksman, $p_3=0.1$, $p_2=0.3$, $p_1=0.6$.

If in a certain firing, 12 shots are fired, then the possible values of the numbers of hits in each of the regions I, II, and III are given by all integers from 0 to 12 inclusive; but this fact in itself still does not enable us to judge the quality of the firing. On the contrary, we can

---

[1] The reader can object saying that hitting region III (i.e., a miss) should not be given a point. However, if a point is given for the right to fire then the one who fired the bad shot has already by the same token received one point.

get a complete picture of this quality if besides the possible values of the number of hits we are also given the probabilities of these values, i.e., the numbers which indicate how frequently in a series of 12 shots one encounters a certain number of hits in each region.

Clearly, the situation in all cases will be the following: knowing the probabilities of the various possible values of the random variables, we will by the same token know how often we ought to expect the appearance of more favorable or less favorable values of them, and this, manifestly, is sufficient to judge the effectiveness or the quality of this operation with which the given random variable is concerned. Practice shows that knowledge of the probabilities of all possible values of the random variable under investigation is in reality sufficient for the solution of any problem connected with the use of this random variable as an index to estimate the quality of the corresponding operation. We thus arrive at the result that, for a complete characterization of a certain random variable as such, it is necessary and sufficient to know:

1) the enumeration of all possible values of this variable and
2) the probability of each of these values.

From this it is clear that it is convenient to specify a random variable by means of a table having two rows: the upper row contains in any order the possible values of the random variable and the lower row contains their probabilities, so that under each of the possible values is placed its probability. Thus, in the example we considered above, the number of points awarded in one shot by the best marksman can, as a random variable, be represented by the table

TABLE I

1	2	3
0	0.2	0.8

In the general case, a random variable whose possible values are $x_1, x_2, \ldots, x_n$ and whose corresponding probabilities are $p_1, p_2, \ldots, p_n$ is represented by the table

x_1	x_2	$\cdots$	x_n
p_1	p_2	$\cdots$	p_n

To give such a table, i.e., to give all possible values of the random variable together with their probabilities, means, as we say, that we give the *distribution law* of this random variable.   Knowing the distribution law of a given random variable enables one to solve all probability problems connected with it.

PROBLEM.   The number of points awarded to a marksman for one shot has the distribution law (I); the same number of points for the second marksman has the following distribution law:

TABLE II

1	2	3
0.2	0.5	0.3

Find the distribution law for the sum of the points awarded both marksmen.

It is clear that the sum we are dealing with here is a random variable; our problem is to set up its table.   To this end, we must consider all possible results of the combined firing of our two marksmen.   We arrange these results in the following table, where the probability of each result is calculated according to the multiplication rule for independent events and where x denotes the number of points awarded the first marksman and y is the number of points awarded the second marksman:

No. of the result	x	y	$x+y$	Probability of the the result
(1)	1	1	2	$0 \cdot 0.2 = 0$
(2)	1	2	3	$0 \cdot 0.5 = 0$
(3)	1	3	4	$0 \cdot 0.3 = 0$
(4)	2	1	3	$0.2 \cdot 0.2 = 0.04$
(5)	2	2	4	$0.2 \cdot 0.5 = 0.1$
(6)	2	3	5	$0.2 \cdot 0.3 = 0.06$
(7)	3	1	4	$0.8 \cdot 0.2 = 0.16$
(8)	3	2	5	$0.8 \cdot 0.5 = 0.4$
(9)	3	3	6	$0.8 \cdot 0.3 = 0.24$

This table shows that the sum $x+y$ which interests us can assume the values 3, 4, 5, and 6; the value 2 is impossible inasmuch as its probability equals zero.[1]   We have $x+y=3$ in the cases of the results (2) and (4); hence, in order that the sum $x+y$ obtain the value 3, it is necessary that one of the results (2) or (4) occur and the probability of this, according to the addition law, equals the sum of the probabilities of these results, i.e., it is equal to $0+0.04=0.04$.   For the sum

$$x+y = 4$$

it is necessary that one of the results (3), (5), or (7) occur; the probability of this sum is therefore equal (again according to the addition rule) to $0+0.1+0.16=0.26$.   In a similar manner, we find that the probability that the sum $x+y$ has the value 5 equals

$$0.06+0.4 = 0.46,$$

and the probability of the value 6, which appears only in the case of result (9), equals 0.24.   Thus, for the random variable $x+y$ we obtain the following table of possible values and their probabilities:

TABLE III

3	4	5	6
0.04	0.26	0.46	0.24

Table III solves the problem posed completely.

The sum of all four probabilities in Table III equals unity; every distribution law should, of course, possess this property inasmuch as we are dealing with the sum of the probabilities of all possible values of a random variable, i.e., with the sum of the probabilities of some complete system of events.   It is convenient to make use of this property of distribution laws as a method for checking the accuracy of the calculations made.

---

[1] One can, of course, consider the number 2 also to be a possible value of the quantity $x+y$ having probability 0.   This is similar to what we did, for the sake of generality, for the value 1 in Table I.

CHAPTER 8

MEAN VALUES

§ 20. Determination of the mean value of a random variable

The two marksmen we were just discussing, when firing together, can earn either 3 or 4 or 5 or 6 points, depending on the random circumstances; the probabilities of these four possible results are indicated in Table III on page 64. If we ask "How many points do the two marksmen earn with one (double) shot?", we are unable to give an answer to this question because different shots yield different results. But, in order to estimate the quality of firing of our two marksmen, we will, of course, be interested in the result not of a single pair of shots (this result can be random) but in the *average* result after an entire series of pairs of shots. But how many points on the average does one pair of shots by our marksmen yield? This question is posed in an altogether intelligible way and a clear-cut answer to it can be given.

We shall reason as follows. If the pair of marksmen shoot a hundred double shots, then, as is shown by Table III

approximately 4 of these shots yield 3 points each
,,   26   ,,   ,,   ,,   4   ,,
,,   46   ,,   ,,   ,,   5   ,,
,,   24   ,,   ,,   ,,   6   ,,

Thus, on the average, each group of one hundred double shots yields the pair of marksmen a total number of points which is expressed by the sum

$$3 \cdot 4 + 4 \cdot 26 + 5 \cdot 46 + 6 \cdot 24 = 490.$$

Dividing this number by 100, we obtain that on the average 4.9 points are awarded for each shot; and this yields the answer to the question we posed. We note that instead of dividing the sum total (490) by 100 (as we have just done), we could, even before adding, have divided each of the terms by 100; then the sum gives us directly the average number of points for one shot. It is simplest to carry out this division by dividing the second factor of each term by 100; for these factors were obtained by multiplying the probabilities, indicated in Table III, by 100, and, therefore, to divide them by 100, it suffices simply to return to these probabilities. For the average number of points awarded for one shot we thus obtain the expression:

$$3 \cdot 0.04 + 4 \cdot 0.26 + 5 \cdot 0.46 + 6 \cdot 0.24 = 4.9.$$

The sum appearing in the left member of this equality, as we see directly, is formed from the data in Table III by a very straightforward rule: each of the possible values indicated in the upper row of this table is multiplied by its probability appearing under it in the table and all such products are then added.

We now employ this same line of reasoning for the general case. We assume that a certain random variable is given by the table

x_1	x_2	$\ldots$	x_k
p_1	p_2	$\ldots$	p_k

We recall that if the probability of the value x_1 of the quantity x equals p_1, then this signifies that in a series of n operations this value x_1 will be observed approximately n_1 times, where $n_1/n = p_1$, which implies that $n_1 = np_1$; analogously, the value x_2 in this connection is encountered approximately $n_2 = np_2$ times, and so on. Thus, a series of n operations will contain, on the average,

$$n_1 = np_1 \text{ operations where } x = x_1$$
$$n_2 = np_2 \quad \text{,,} \qquad \text{,,} \quad x = x_2$$
$$\cdot \quad \cdot \quad \cdot \quad \cdot \quad \cdot \quad \cdot \quad \cdot \quad \cdot$$
$$n_k = np_k \quad \text{,,} \qquad \text{,,} \quad x = x_k.$$

The sum of the values of the quantity x in all n operations carried out will therefore be approximately equal to

$$x_1 n_1 + x_2 n_2 + \ldots + x_k n_k = n(x_1 p_1 + x_2 p_2 + \ldots + x_k p_k).$$

Therefore, the mean value $\bar{x}$ of a random variable, corresponding to an individual operation and obtained from the sum just written down by dividing by the number n of operations in the given series, will be equal to

$$\bar{x} = x_1 p_1 + x_2 p_2 + \ldots + x_k p_k.$$

We thus arrive at the following important rule: *to obtain the mean value (or mathematical expectation or expected value) of a random variable we must multiply each of its possible values by the corresponding probability and then add up all the products obtained.*

Of what benefit to us can knowledge of the mean value of a random variable be? In order to answer this question more convincingly, we shall first consider a few examples.

EXAMPLE 1. We return once more to the two marksmen. The number of points they are awarded are random variables whose distribution laws are given by Table I for the first marksman and by Table II for the second (see pages 62 and 63). One careful glance at these two tables already shows us that the first shoots better than the second; in fact, the probability of the best result (3 points) is significantly greater for him than for the second marksman whereas, in contrast, the probability of the worst result is greater for the second marksman than for the first. However, such a comparison is not satisfactory as it is purely qualitative in character—we still do not possess that measure or that number whose magnitude would give a direct estimate of the quality of the firing of one or the other marksman in a way similar to the way in which the temperature, for instance, directly estimates the amount of heating of a physical body. Not having such an estimation measure, we can always encounter the case for which a direct consideration does not yield an answer or for which this answer can be questionable. Thus, if, instead of Tables I and II, we had the tables

TABLE I'

1	2	3
0.4	0.1	0.5

for the first marksman

TABLE II'

1	2	3
0.1	0.6	0.3

for the second marksman

then it would be difficult by a single glance at these tables to decide which of the two marksmen shoots better; true—the best result (3 points) is more probable for the first than for the second, but at the same time the worst result (1 point) is also more probable for him than for the second; in contrast, the result of 2 points is more probable for the second than for the first.

We now form, by the rule indicated above, the mean value of the number of points for each of our two marksmen:

1) for the first marksman,

$$1 \cdot 0.4 + 2 \cdot 0.1 + 3 \cdot 0.5 = 2.1;$$

2) for the second marksman,

$$1 \cdot 0.1 + 2 \cdot 0.6 + 3 \cdot 0.3 = 2.2.$$

We see that the second marksman wins on the average a slightly greater number of points than the first; in practice, this means that in a repeated firing the second marksman will, generally speaking, produce a somewhat better result than the first. We now say with conviction that the second marksman shoots better. The mean value of the number of points won gave us a suitable measure with the aid of which we can easily, and by a method which leaves no doubt, compare the skills of the different marksmen with one another.

EXAMPLE 2. In assembling a precision instrument, it might be required, depending on one's success, to make 1, 2, 3, 4, or 5 trials for the exact fitting of a certain part. Thus, the number x of trials necessary to attain a satisfactory assembly is a random variable with the possible values 1, 2, 3, 4, 5; suppose the probabilities of these values are given by the following table:

1	2	3	4	5
0.07	0.16	0.55	0.21	0.01

We can study the problem of supplying a given assembler with the number of parts needed for 20 instruments.[1] In order to be able to make an approximative estimate of this number, we cannot make use of the given table directly—it tells us only that in various cases the situation is different. But, if we find the mean value $\bar{x}$ of the number of trials x which are necessary for one instrument and multiply this mean value by 20, then we obviously obtain an approximate value of the number sought. We find that

$$\bar{x} = 1 \cdot 0.07 + 2 \cdot 0.16 + 3 \cdot 0.55 + 4 \cdot 0.21 + 5 \cdot 0.01 = 2.93;$$
$$20\bar{x} = 2.93 \cdot 20 = 58.6 \approx 59.$$

In order that the assembler have a small surplus to take care of the case when the expenditure of parts actually exceeds expectation, it will be useful in practice to give him 60–65 parts.

In the examples considered, we are dealing with a situation in which for a certain random variable practice requires a known initial, approximate estimate; we cannot give such an estimate by a single glance at the table—the table tells us only that our random variable

---

[1] In this connection, we will assume that a part rejected in assembling one instrument is no longer used in assembling others.

can assume such-and-such values with such-and-such probabilities. But the *mean value* of the random variable calculated by this table is already capable of yielding such an estimate because this is namely the value which our quantity will assume on the average in a more or less extended series of operations. We see that from the practical side, the mean value characterizes the random variable especially well when we are dealing with a mass operation or one that is repeated frequently.

PROBLEM 1.   A series of trials with the same probability p of the occurrence of a certain event A is carried out, in which connection the results of the individual trials are mutually independent. Find the mean value of the number of occurrences of the event A in a series of n trials.

The number of occurrences of event A in a series of n trials is a random variable with possible values $0, 1, 2, \ldots, n$, where the probability of the value k is equal, as we know, to

$$P_n(k) = \frac{n!}{k!(n-k)!} p^k (1-p)^{n-k}.$$

Therefore, the mean value sought equals

$$\sum_{k=0}^{n} kP_n(k).$$

We calculated this sum in the course of the proof of Bernoulli's theorem (see page 53) and we saw that it was equal to np. We also verified that the *most probable* number of occurrences of the event A in n trials is, in the case of large n, close to np. We now see that the *average* number of occurrences of the event A for arbitrary n is precisely equal to np. Thus, in the given case, the most probable value of the random variable coincides with its mean value. We must, however, avoid thinking that this coincidence holds for arbitrary random variables for, in general, the most probable value of a random variable can be very far removed from its mean value. Thus, for instance, for the random variable with distribution law

0	5	10
0.7	0.1	0.2

the most probable value is 0 and the mean value is 2.5.

PROBLEM 2. Independent trials are made in each of which some event A can occur with probability 0.8. Trials are made until the event A occurs; the total number of trials does not exceed four. Determine the average number of trials made.

The number of trials which are to be made under the conditions of the problem can equal 1, 2, 3, or 4. We must calculate the probability of each of these four values. In order that only one trial suffice, it is necessary that the event A occur at the first trial; the probability of this is

$$p_1 = 0.8.$$

In order that exactly two trials be required, it is necessary that the event A does not occur at the first trial and that it does occur at the second trial. The probability of this, by the multiplication rule for independent events, equals

$$p_2 = (1-0.8)\cdot 0.8 = 0.16.$$

In order that three trials be required, it is necessary that the event A does not occur at the first two trials but that A occurs at the third trial. Therefore,

$$p_3 = (1-0.8)^2\cdot 0.8 = 0.032.$$

Finally, the necessity for four trials arises under the condition that A does not occur for the first three trials (independently of what the fourth trial yields); therefore

$$p_4 = (1-0.8)^3 = 0.008.$$

Thus, the number of trials made, as a random variable, is determined by the distribution law

1	2	3	4
0.8	0.16	0.032	0.008

The mean value of this number therefore equals

$$1\cdot 0.8 + 2\cdot 0.16 + 3\cdot 0.032 + 4\cdot 0.008 = 1.248.$$

If, for instance, 100 such observations are to be made, then it can be assumed that approximately $1.248\cdot 100 \approx 125$ trials will have to be made.

In practice one frequently encounters problems formulated this way. For example, we test a yarn for strength and we give it a higher classification if it does not break even once under the load P when samples of standard length from the same skein (or lot) are tested. Each time no more than four samples are tested.

PROBLEM 3.   A certain plot of ground has the form of a square whose side according to given aerial photographic measurements equals 350 m.   The quality of the aerial photograph is determined by the fact that an error of

$$
\begin{array}{lll}
\text{0 m. has probability 0.42} \\
\pm 10 \text{ m.} & \text{,,} & 0.16* \\
\pm 20 \text{ m.} & \text{,,} & 0.08 \\
\pm 30 \text{ m.} & \text{,,} & 0.05.
\end{array}
$$

Find the mean value of the area of the plot.

Depending on the randomness of the aerial photographic measurement, the side of the plot is a random variable whose distribution law is given by the table

TABLE I

320	330	340	350	360	370	380
0.05	0.08	0.16	0.42	0.16	0.08	0.05

From this we can at once find the mean value of this random variable, since in the given case we do not even need to apply our computation rule; in fact, since the same errors in one or another direction are equally probable, it is already clear from symmetry that the mean value of the side of the plot equals the observed value, i.e., 350 m.   In more detail, the expression for the mean value will contain the terms

$$
\begin{aligned}
(340+360)\cdot 0.16 &= [(350+10)+(350-10)]\cdot 0.16 \\
&= 2\cdot 350\cdot 0.16 \\
(330+370)\cdot 0.08 &= 2\cdot 350\cdot 0.08 \\
(320+380)\cdot 0.05 &= 2\cdot 350\cdot 0.05;
\end{aligned}
$$

it is therefore equal to $350\cdot(0.42+2\cdot 0.16+2\cdot 0.08+2\cdot 0.05)=350$.

One might conclude that from these same symmetry considerations, the mean value of the *area* of the plot must equal $350^2 = 122,500$ m.^{2}; this would be the case if the mean value of the square of a random

---

* This is understood to mean that the error $+10$ m. and the error -10 m. each have the probability 0.16; the same is to be understood for the other probabilities.

variable equalled the square of its mean value. This is, however, not
the case; in our example, the area of the plot can have the values

$$320^2, 330^2, 340^2, 350^2, 360^2, 370^2, 380^2.$$

Now which of these values holds in reality depends on which of the
seven cases listed in Table I is present, so that the probabilities of these
seven values are the same as the probabilities in Table I; more briefly,
the distribution law of the area of the plot is given by the table

320^2	330^2	340^2	350^2	360^2	370^2	380^2
0.05	0.08	0.16	0.42	0.16	0.08	0.05

and, consequently, its mean value equals

$$320^2 \cdot 0.05 + 330^2 \cdot 0.08 + 340^2 \cdot 0.16 + 350^2 \cdot 0.42 + 360^2 \cdot 0.16$$
$$+ 370^2 \cdot 0.08 + 380^2 \cdot 0.05.$$

And, here, it is useful in order to shorten the calculations to make use
of the symmetry on hand; we must see how this is done because such
possibilities of simplification arise rather frequently. We can rewrite
the above expression in the form

$$350^2 \cdot 0.42 + (340^2 + 360^2) \cdot 0.16 + (330^2 + 370^2) \cdot 0.08$$
$$+ (320^2 + 380^2) \cdot 0.05$$

$$= 350^2 \cdot 0.42 + [(350 - 10)^2 + (350 + 10)^2] \cdot 0.16$$
$$+ [(350 - 20)^2 + (350 + 20)^2] \cdot 0.08$$
$$+ [(350 - 30)^2 + (350 + 30)^2] \cdot 0.05$$

$$= 350^2 [0.42 + 2 \cdot 0.16 + 2 \cdot 0.08 + 2 \cdot 0.05]$$
$$+ 2 \cdot 10^2 \cdot 0.16 + 2 \cdot 20^2 \cdot 0.08 + 2 \cdot 30^2 \cdot 0.05$$

$$= 350^2 + 2 \cdot (16 + 32 + 45) = 122{,}686.$$

In this method of calculation, all computations can be made "mentally."

We see that the mean value of the areas of the plot turned out to be
somewhat larger (it is true that in practice the difference in this case is
imperceptible) than the square of the mean value of a side (i.e.,
larger than $350^2 = 122{,}500$). It is easily proved that at the base of
this lies a general rule: the mean value of the square of an arbitrary
random variable is always larger than the square of its mean value.

In fact, suppose we have a random variable x with a perfectly arbitrary distribution law

x_1	x_2	$\ldots$	x_k
p_1	p_2	$\ldots$	p_k

then the quantity x^2 will have the following distribution law:

x_1^2	x_2^2	$\ldots$	x_k^2
p_1	p_2	$\ldots$	p_k

The mean values of these two quantities are equal to

$$\bar{x} = x_1 p_1 + x_2 p_2 + \ldots + x_k p_k$$

and

$$\overline{x^2} = x_1^2 p_1 + x_2^2 p_2 + \ldots + x_k^2 p_k,$$

respectively.   We have

$$\overline{x^2} - (\bar{x})^2 = \overline{x^2} - 2(\bar{x})^2 + (\bar{x})^2. \qquad (*)$$

Since $p_1 + p_2 + \ldots + p_k = 1$, the three terms in the right member of (*) can be written in the form

$$\overline{x^2} = \sum_{i=1}^{k} x_i^2 p_i,$$

$$2(\bar{x})^2 = 2(\bar{x})(\bar{x}) = 2\bar{x} \sum_{i=1}^{k} x_i p_i = \sum_{i=1}^{k} 2\bar{x} x_i p_i,$$

$$(\bar{x})^2 = (\bar{x})^2 \sum_{i=1}^{k} p_i = \sum_{i=1}^{k} (\bar{x})^2 p_i,$$

respectively; therefore

$$\overline{x^2} - (\bar{x})^2 = \sum_{i=1}^{k} \{x_i^2 - 2\bar{x} x_i + (\bar{x})^2\} p_i = \sum_{i=1}^{k} (x_i - \bar{x})^2 p_i.$$

Since all terms in the sum in the right member are non-negative, we have

$$\overline{x^2} - (\bar{x})^2 > 0,$$

which was to be proved.

CHAPTER 9

MEAN VALUE OF A SUM AND OF A PRODUCT

§ 21. Theorem on the mean value of a sum

We frequently must calculate the mean value of the sum of two random variables (and sometimes also of a larger number) whose mean values (i.e., mathematical expectations) are known. Suppose, for instance, that two factories manufacture the same product, where it is known that on the average the first factory produces 120 articles daily and the second 180. Can we, with the aid of these data, establish the mean value of the number of articles which one could expect daily from both factories together? Or are these data insufficient and must we know, besides the mean values, something more about the two random variables under consideration (for instance, must we know their distribution laws completely)?

It is very important that for the calculation of the mean value of a sum it be sufficient in all cases to know the mean values of the summands, and that the mean value of the sum be expressed in all cases in terms of the mean values of the summands in the very simplest manner which one could possibly imagine: *the mean value of a sum always equals the sum of the mean values of the summands.* Thus, if x and y are two perfectly arbitrary random variables, then

$$\overline{x+y} = \bar{x}+\bar{y}.$$

In the example introduced above x is the number of articles of the first factory and y is the number of articles of the second factory: $\bar{x}=120$, $\bar{y}=180$ and, hence,

$$\overline{x+y} = \bar{x}+\bar{y} = 300.$$

In order to prove the asserted rule in the general case, we shall assume that the quantities x and y, respectively, are subject to the following distribution laws.

TABLE I

x_1	x_2	$\cdots$	x_k
p_1	p_2	$\cdots$	p_k

TABLE II

y_1	y_2	$\cdots$	y_l
q_1	q_2	$\cdots$	q_l

Then, the possible values of the quantities $x+y$ will be all possible sums of the form x_i+y_j, where $1 \le i \le k$ and $1 \le j \le l$. The probability of the value x_i+y_j, which we shall denote by p_{ij}, is unknown; this is the probability of the two-fold event $x=x_i$, $y=y_j$ (i.e., the probability that the quantity x will have the value x_i and that the quantity y will have the value y_j). If we could assume these two events to be mutually independent, then by the multiplication rule we would, of course, have

$$p_{ij} = p_i q_j, \tag{1}$$

but from this point on we shall no longer assume these events to be independent. Thus, equality (1), generally speaking, will not hold, and we must take into consideration that knowledge of Tables I and II does not permit us to conclude anything about the quantities p_{ij}.

By the general rule, the mean value of the quantity $x+y$ equals the sum of the products of all possible values of this quantity by the corresponding probabilities, i.e.,

$$
\begin{aligned}
\overline{x+y} &= \sum_{i=1}^{k} \sum_{j=1}^{l} (x_i+y_j) p_{ij} \\
&= \sum_{i=1}^{k} x_i \left(\sum_{j=1}^{l} p_{ij} \right) + \sum_{j=1}^{l} y_j \left(\sum_{i=1}^{k} p_{ij} \right).
\end{aligned} \tag{2}
$$

We now consider more carefully the sum $\sum_{j=1}^{l} p_{ij}$; this is the sum of the probabilities of all possible events of the form $(x=x_i, y=y_j)$, where the number i is the same in all terms of the sum and the number j varies from term to term, ranging over all its possible values from 1 to l inclusive. Since the events $y=y_j$ are obviously incompatible for distinct j's, then, by the addition rule, the sum $\sum_{j=1}^{l} p_{ij}$ is the probability of the occurrence of *any one of the l events* $(x=x_i, y=y_j)$ $(j=1, 2, \ldots, l)$.

But to say "some one of the events $x=x_i$, $y=y_j$ $(1 \le j \le l)$ occurred" is entirely equivalent to simply saying "the event $x=x_i$ occurred"; in fact: 1) if one of the events $(x=x_i, y=y_j)$ (it being immaterial what j is) occurred, then, obviously, the event $x=x_i$ also occurred; 2) conversely, if the event $x = x_i$ occurred, then, inasmuch as y necessarily assumes one of its possible values $y_1, y_2, \ldots, y_j$, some one of the events $(x = x_i, \; y = y_j) \; (1 \leq j \leq l)$ must also occur. Thus, $\sum_{j=1}^{l} p_{ij}$, being the probability of occurrence of any one of the events $(x = x_i, \; y = y_j)$ $(1 \leq j \leq l)$, simply equals the probability of the event $x = x_i$, i.e.,

$$\sum_{j=1}^{l} p_{ij} = p_i.$$

In a perfectly analogous way, we can of course convince ourselves that

$$\sum_{i=1}^{k} p_{ij} = q_j;$$

and substituting these expressions into equality (2), we find that

$$\overline{x+y} = \sum_{i=1}^{k} x_i p_i + \sum_{j=1}^{l} y_j q_j = \bar{x} + \bar{y},$$

which was to be proved.

The theorem we just proved for the case of two terms automatically generalizes to the case of three and more terms; in fact, by virtue of what we just proved, we can write

$$\overline{x+y+z} = \overline{\overline{x+y}+z} = \bar{x} + \bar{y} + \bar{z},$$

and so on.

EXAMPLE. n machines are set up in a certain plant and one article is collected from each machine. Determine the average number of rejected articles if it is known that the probability of producing a reject is p_1 for the first machine, p_2 for the second machine, $\ldots$, p_n for the nth machine.

The number of rejects when analyzing one article is a random variable which is capable of assuming only two values: 1 if this article is a reject and 0 if it is usable. The probabilities of these values for the first machine equal p_1 and $1 - p_1$ respectively, as a consequence of which the average number of rejected articles from the number taken from the first machine equals

$$1 \cdot p_1 + 0 \cdot (1 - p_1) = p_1.$$

For the second machine the average number of rejected articles from among those taken equals p_2, and so forth. The total number of rejected articles is the sum of the rejected articles among the articles produced on the first, second, and the other machines. Therefore, by virtue of the rule for the addition of mean values which we just established, the average number of rejected articles among those chosen equals

$$p_1+p_2+\ldots+p_n,$$

which solves the problem posed.

In particular, if the probability of producing a reject is the same for all machines ($p_1 = p_2 = \ldots = p_n = p$), then the mean value of the total number of rejects equals np. We already obtained this result on page 53 [formula (5)]. It is interesting to compare the cumbersome calculations which we needed for this purpose with the simple line of reasoning, requiring no calculations whatsoever, which led us here to the same result. However, we gained not only in simplicity but also in generality. In our previous derivation, we assumed the results of producing individual articles to be mutually independent events and in reality our former method of derivation was suitable only under this hypothesis; but now we can do without this assumption, since the law of addition of mean values, on which we based our new derivation, holds for arbitrary random variables without any restriction. Thus, whatever the mutual dependence between individual machines and the articles made by them is, if only the probability p of producing a reject is the same for all machines then the mean value of the number of rejected articles is always equal to np for n articles chosen at random.

§ 22. Theorem on the mean value of a product

The same problem, which we solved for a sum of random variables, must frequently also be considered for their products. Suppose the random variables x and y are again subject to the distribution laws indicated by Tables I and II, respectively. Then the product xy is a random variable for which products of the form x_iy_j ($1 \leq i \leq k$, $1 \leq j \leq l$) serve as possible values; the probability of the value x_iy_j equals p_{ij}. The problem consists in finding a rule which would enable us to express, in all cases, the mean value $\overline{xy}$ of the quantity xy in terms of the mean values of the factors. The solution of this problem in the general case, however, turns out to be impossible. The quantity $\overline{xy}$, generally speaking, is not uniquely determined by knowing the mean values $\bar{x}$ and $\bar{y}$ (i.e., various values of the quantity $\overline{xy}$ are possible for the same $\bar{x}$ and $\bar{y}$); as a result of this no general formula can exist which expresses $\overline{xy}$ in terms of $\bar{x}$ and $\bar{y}$.

But there is one particular case when such an expression is possible and then the connection obtained is of an extraordinarily simple

character. We shall agree to call the random variables x and y
mutually independent if the events $x=x_i$ and $y=y_j$ are mutually independent for arbitrary i and j, i.e., if the condition that one of our two
random variables take on one or another definite value does not
influence the distribution law of the second random variable.   If the
quantities x and y are mutually independent in the sense just defined,
then

$$p_{ij} = p_i q_j \quad (i = 1, 2, \ldots, k; \; j = 1, 2, \ldots, l),$$

according to the rule of multiplication of independent events; therefore,

$$\overline{xy} = \sum_{i=1}^{k} \sum_{j=1}^{l} x_i y_j p_{ij} = \sum_{i=1}^{k} \sum_{j=1}^{l} x_i y_j p_i q_j$$

$$= \sum_{i=1}^{k} x_i p_i \sum_{j=1}^{l} y_j q_j = \bar{x} \cdot \bar{y}.$$

*Therefore, for mutually independent random variables, the mean value of the
product equals the product of the mean values of the factors.*

As in the case of addition, this rule which we derived for the product
of two random variables automatically generalizes to the product of an
arbitrary number of factors; in this connection, it is only necessary
that these factors be mutually independent, i.e., that knowledge of any
definite values for each group of these quantities does not influence
the distribution laws of the remaining quantities.

EXAMPLE 1.   We shall assume that it is necessary to measure an
area of rectangular form by means of an aerial photo survey and that
the measurement of the sides of this rectangle give 72 m. and 50 m.
The distribution law of the measurement errors is not known but it is
known that the errors of the same magnitude in one or the other
direction are equally probable; it is then clear from symmetry
considerations (and it can easily be proved—see Problem 3 on page
70) that the mean values of the sides of the rectangle coincide with
the obtained results of measurement.   If these two results of measurement can be considered mutually independent random variables, then
the mean value of the area according to the multiplication rule which
we just derived will be equal to the product of the mean values of its
sides, i.e., $72 \times 50 = 3600$ m.^{2}.   But there can sometimes be a basis
for assuming the measurements of the sides to be mutually dependent.
This will be so, for example, in the case when both measurements are
made with the same improperly calibrated instrument.   If measurement of the length yields a result which significantly exceeds the true

length, then we naturally have the right to assume that the measuring instrument is as a rule inclined to give quantities which are too large, as a consequence of which the probability of exaggerated values will also increase in measurements of the width, so that it is impossible to consider these two quantities to be mutually independent. In such cases, the mean value of the area cannot be taken equal to the product of the mean values of the sides of the rectangle and, to determine it, supplementary information is required.

EXAMPLE 2. Along a conductor, whose resistance depends on random circumstances, there flows an electric current whose strength also depends on chance. It is known that the mean value of the resistance of the conductor equals 25 ohms and that the mean strength of the current equals 6 amperes. It is required that one compute the mean value of the electromotive force (i.e., voltage) E of the current flowing in the conductor.

According to Ohm's law,

$$E = RI,$$

where R is the resistance of the conductor and I is the strength of the current. According to our assumption,

$$\bar{R} = 25, \qquad \bar{I} = 6;$$

then, assuming the quantities R and I to be mutually independent, we find that

$$\bar{E} = \bar{R} \cdot \bar{I} = 25 \cdot 6 = 150 \text{ volts.}$$

CHAPTER 10

DISPERSION AND MEAN DEVIATIONS

§ 23. Insufficiency of the mean value for the characterization of a random variable

We have already seen repeatedly that the mean value of a random variable gives us an approximate, initial "measure" of the variable and that there are many cases when, for the practical purposes confronting us, this representation is sufficient. Thus, for the comparison of the proficiency of two marksmen in a competition it is sufficient for us to know the mean value of the number of points won by them; for the comparison of the effectiveness of two different ways of computing the number of cosmic particles it is completely sufficient to know the mean value of the number of particles not counted which these two systems are capable of admitting; and so on. In all these cases, we gain an essential advantage by describing our random variable by one number—its mean value—instead of giving it by a complicated distribution law. The situation appears as though we had before us not a random variable but rather a quantity, with a perfectly well-defined value, which is known with certainty.

However, much more frequently we encounter another state of affairs in which the features of a random variable which are most important for practical purposes are not characterized by its mean value to any extent whatsoever, but require a more detailed knowledge of its law of distribution. We have a typical case of this sort in the investigation of the distribution of errors in measurement. Let x be the magnitude of the error, i.e. the deviation of the value obtained of the quantity being measured from its true value. In the absence of systematic errors the mean value of the errors of measurement, which we shall denote by $\bar{x}$, equals zero. We shall assume that the measurements are carried out under this condition. The question arises, how will the errors be distributed? How frequently will an error of a given magnitude be encountered? Knowing only the value $\bar{x}=0$, we cannot obtain any answer to all these questions. We know only that positive and negative errors are possible and that their chances are approximately the same because the mean value of the magnitudes of the errors equals zero.   But we do not know the most important thing: will the results of measurement—in the majority of cases—lie close to the true value of the quantity being measured so that we can count on each measurement result with a high degree of certainty, or will they mostly be scattered at great distances in both directions from the true measurement? Both possibilities are completely admissible.

Two observers, making measurements with the same mean value of error $\bar{x}$, can obtain measurements of different degrees of precision. It can happen that one of them yields systematically a greater "dispersion" of the measurement results than the other.   This means that for this observer the errors can take on larger values on the average and hence the measurements will deviate more from the quantity being measured than for the other observer.   And this is possible although the mean value of the measurement errors is the same for both observers.

Let us consider another example. Let us imagine that two varieties of wheat are tested for productivity. Depending on the random circumstances (e.g., the amount of rainfall, the distribution of fertilizer, the amount of sunlight, etc.), the harvest from a square meter is subject to significant variations and represents a random variable.   Let us assume that under the same conditions the average harvest is the same for every variety—for instance, 240 grams per square meter.   Can one judge the quality of the variety being tested only by knowing the value of an average harvest?   Clearly not, since that variety is of the greatest economic interest whose productivity is least subject to the random influences of weather and other factors—in other words, for which the "dispersion" of productivity is the least. We thus see that in testing one variety of wheat against another for productivity the extent of its possible variations has an importance, no less than the average productivity.

§ 24. Various methods of measuring the dispersion of a random variable

The examples introduced above, as well as a number of others which are analogous to them, show convincingly that in many cases in order to describe certain interesting features of random variables, the specification of their mean values is simply insufficient. These features, which are of practical interest, remain completely unknown for such specifications, and to describe them we must either have before us the entire distribution table, which is almost always complicated and inconvenient, or else we endeavor to introduce for the desired description, besides the mean value, one or two numbers of a similar type so that this small group of numbers gives a sufficient practical characterization of those features of the quantity studied which are of importance to us. We now consider how this last possibility can be realized.

As the examples which we have considered show, the most important question in many practical cases turns out to be how large, generally speaking, the deviations of the values which are actually assumed by the given random variable are from their mean value, i.e., how extensively these values are strewn, scattered, or dispersed. Will they be, for the most part, closely grouped around a mean value (and hence also among themselves), or, on the contrary, will the majority of them differ very markedly from the mean value (in which case, certain of them will, of necessity, also differ significantly from one another)?

The following crude scheme enables us to form a clear picture of this difference. We consider two random variables with the following probability distributions:

TABLE I

-0.01	$+0.01$
0.5	0.5

TABLE II

-100	$+100$
0.5	0.5

Both random variables whose tables we have shown have zero for their mean value, but, whereas the first of them always assumes values very close to zero (and close to one another), the second, in contrast, is capable of assuming only values which differ considerably from zero (and from one another). For the first quantity, knowing its mean value gives us, at the same time, initial information of all its actual possible values; but for the second, the mean value is removed very significantly from the actual possible values and does not give any representation of them. We say, in the second case, that the possible values are *dispersed* much more than in the first.

Thus, our problem is to find a number which could, in an intelligible way, give us a *measure* of the dispersion of the random variable which would at least indicate how large we must expect deviations of this quantity from its mean value to be. The deviation $x - \bar{x}$ of a random variable from its mean value $\bar{x}$ is itself obviously a random variable;

the absolute value $|x - \bar{x}|$ of this deviation, which characterizes its magnitude without dependence on sign, is also a random variable. It is desirable to have a number which could basically characterize its random deviation $|x - \bar{x}|$—to tell us how large, for instance, this deviation can turn out to be.   To solve this problem, there exist many different methods, of which the following three are the most frequently used in practice.

1. *Mean deviation.*   For the first evaluation of the random variable $|x - \bar{x}|$ it is most natural to take its mean value $\overline{|x - \bar{x}|}$.   This mean value of the absolute value of the deviation is called the *mean deviation* of the quantity x.   If the random variable x is given by the table

x_1	x_2	$\ldots$	x_k
p_1	p_2	$\ldots$	p_k

then the table for the random variable $|x - \bar{x}|$ has the form

$|x_1 - \bar{x}|$	$|x_2 - \bar{x}|$	$\ldots$	$|x_k - \bar{x}|$
p_1	p_2	$\ldots$	p_k

where $\bar{x} = \sum\limits_{i=1}^{k} x_i p_i$.   For the mean deviation M_x of the quantity x, we obtain the formula

$$M_x = \overline{|x - \bar{x}|} = \sum_{i=1}^{k} |x_i - \bar{x}| p_i,$$

where, of course, we again have $\bar{x} = \sum\limits_{i=1}^{k} x_i p_i$.   For quantities given by Tables I and II, $\bar{x} = 0$, and we have

$$M_{x_I} = 0.01 \quad \text{and} \quad M_{x_{II}} = 100,$$

respectively.   But both examples are trivial since in both cases the absolute value of the deviation turns out to be capable of assuming only one value, thus losing its random variable character.

We calculate, further, the mean deviation for random variables, defined by Tables I' and II' on page 67.   We saw there that the mean values of these quantities are equal to 2.1 and 2.2, respectively; i.e., they are very close to one another.   The mean deviation for the first quantity equals

$$|1-2.1|\cdot 0.4 + |2-2.1|\cdot 0.1 + |3-2.1|\cdot 0.5 = 0.90,$$

and for the second it is

$$|1-2.2|\cdot 0.1 + |2-2.2|\cdot 0.6 + |3-2.2|\cdot 0.3 = 0.48.$$

We see that the mean deviation for the second quantity is just about half as large as for the first.   In practice this means, obviously, that although both marksmen win, on the average, approximately the same number of points—and in this sense they can be acknowledged to be equally skillful—for the second marksman the firing is of a significantly *more uniform* nature.   His results are significantly less dispersed than those of the first marksman who, with the same average number of points, fires unevenly, frequently giving results which are much better as well as much worse than the average.

2. *Standard deviation.*   It is very natural to measure initially the magnitude of the dispersion with the aid of the mean deviation, but at the same time it is also very inconvenient in practice, inasmuch as calculations and estimations involving absolute values are frequently complicated and sometimes even completely unfeasible.   Therefore, in practice, we usually prefer to introduce another measure for the magnitude of dispersion.

Like the deviation $x-\bar{x}$ of the random variable x from its mean value $\bar{x}$, the square $(x-\bar{x})^2$ of this deviation is also a random variable, whose table in our old notation has the form

$(x_1-\bar{x})^2$	$(x_2-\bar{x})^2$	$\ldots$	$(x_k-\bar{x})^2$
p_1	p_2	$\ldots$	p_k

Hence the mean value of this square equals

$$\sum_{i=1}^{k} (x_i-\bar{x})^2 p_i.$$

This quantity gives us an idea of what the *square* of the deviation $x-\bar{x}$ is approximately equal to.   Extracting the square root of this quantity, i.e.,

$$Q_x = \sqrt{\sum_{i=1}^{k} (x_i-\bar{x})^2 p_i},$$

we obtain a quantity capable of characterizing for us the approximate magnitude of the deviation itself. The quantity Q_x is called the *standard deviation* (or the *root mean square deviation*) of the random variable x, and its square Q_x^2 is called its *variance*. Of course, this new measure of the magnitude of the dispersion is of a somewhat more artificial character than the mean deviation which we introduced above. Here we go along a roundabout way, first finding an approximate value for the *square* of the standard deviation (i.e., for the variance) and then, by extracting the square root, returning to the standard deviation itself. But, in compensation, as we shall see in the next section, the utilization of the standard deviation Q_x significantly simplifies the calculations. It is this, namely, which forces statisticians to prefer to make use of the standard deviation in practice.

EXAMPLE. For the random variables defined by Tables I' and II' on page 67, we have

$$Q_{x_{\mathrm{I}'}}^2 = (1-2.1)^2 \cdot 0.4 + (2-2.1)^2 \cdot 0.1 + (3-2.1)^2 \cdot 0.5 = 0.89$$

and

$$Q_{x_{\mathrm{II}'}}^2 = (1-2.2)^2 \cdot 0.1 + (2-2.2)^2 \cdot 0.6 + (3-2.2)^2 \cdot 0.3 = 0.36,$$

respectively; consequently,

$$Q_{x_{\mathrm{I}'}} = \sqrt{0.89} \approx 0.94 \quad \text{and} \quad Q_{x_{\mathrm{II}'}} = 0.60.$$

Previously, for these same random variables we found the mean deviations to be

$$M_{x_{\mathrm{I}'}} = 0.90 \quad \text{and} \quad M_{x_{\mathrm{II}'}} = 0.48.$$

We see that the standard deviation, as well as the mean deviation, is significantly larger for the first random variable than for the second; whether we measure the dispersion by the mean deviation or by the standard deviation, in either case we arrive at the same result: the first of our two random variables is dispersed to a greater degree than the second.

In each case before us the standard deviation turned out to be greater than the mean deviation; it is easy to see that this is what should be the case for an arbitrary random variable. In fact, the variance Q_x^2, as the mean value of the square of the quantity $|x-\bar{x}|$, according to the rule proved on page 73, cannot be less than the square of the mean value M_x of the quantity $|x-\bar{x}|$, and it follows from $Q_x^2 \geq M_x^2$ that $Q_x \geq M_x$.

3. *Probable (or equally likely) deviation.* Frequently, especially in military operations, another method is utilized for determining the extent of dispersions; we shall discuss it in terms of an artillery example.

We assume that artillery fire is executed from the point O in the direction OX (Fig. 9); the distance x of the spot where the shell hits from the place of firing is a random variable whose mean value $\bar{x}$ indicates to us the position of the "center of impact" C $(OC = \bar{x})$. The points of impact of the individual shells are more or less dispersed about C.

The deviation $x - \bar{x}$ of the random variable we are studying (i.e., the range of the shell) from its mean value is at the same time the deviation of the point of impact of the shell from the center of impact C; every estimate of the quantity $|x - \bar{x}|$ therefore estimates at the same time the degree of dispersion—of scattering—of the missiles about this center C and thus serves as an important index of the quality of fire.

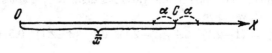

FIG. 9

If we mark off a very small segment of length α to the left and right of the point C, then only a small number of shells will fall inside the segment of length 2α with center at the point C, obtained in this way —in other words, the probability that $|x - \bar{x}| < \alpha$ will still be small for small α. But we shall now extend the segment we constructed by increasing the number α (which, as we know, was chosen arbitrarily). The larger the segment constructed, the larger will be the fractional part of the shells falling inside it and correspondingly, the larger will be the probability of an individual shell falling inside this segment. When α is very large, then practically all the shells will fall inside this segment; thus, with the continuous increase of the number α, the probability of the inequality

$$|x - \bar{x}| < \alpha$$

increases from zero to one.   At first, for small α, it is more probable that

$$|x - \bar{x}| > \alpha,$$

i.e., that the shell will fall outside the segment; and then for large α, it is more probable that we will have $|x - \bar{x}| < \alpha$, i.e., that the shell will fall inside the segment.   Therefore, somewhere in the transition from small values of the number α to large ones, there must be a value α_0 of this number α such that the shell has equal probability of falling inside or outside the segment of length $2\alpha_0$, constructed about the point C. In other words, the inequalities

$$|x - \bar{x}| < \alpha_0$$

and

$$|x - \bar{x}| > \alpha_0$$

are equally probable and hence the probability of each of them equals 1/2 (if we agree to disregard the negligibly small probability of the exact equality $|x - \bar{x}| = \alpha$). For $\alpha < \alpha_0$, the second inequality written down above is the more probable and for $\alpha > \alpha_0$ the first. Thus, there exists a uniquely defined number α_0 such that the absolute value of the deviation can turn out to be larger or smaller than α_0 with equal probability.

How large α_0 is depends on the qualities of the cannon being fired. It is easily seen that the quantity α_0 can serve as a measure of dispersion of the shells, in a manner similar to the mean deviation or the standard deviation. In fact, if, for example, α_0 is very small, then this means that half of all shells fired by the first cannon already fall in a very small segment surrounding the point C, which implies that there is comparatively insignificant dispersion. In contrast, if α_0 is large, then, even after surrounding the point C with a large segment, we must, in spite of everything, consider that half of the shells will fall outside the bounds of this segment; this evidently shows that the shells are widely dispersed about the center.

The number α_0 is usually called the *equally likely* (or *probable*) *deviation* of the quantity x; thus, the probable deviation of the random variable x is the number such that the deviation $x - \bar{x}$ can turn out to be in absolute value larger as well as smaller than this number with equal probability. Although the probable deviation of the quantity x which we shall denote by E_x is not more convenient, for mathematical calculations, than the mean deviation M_x and much less convenient than the standard deviation Q_x, nevertheless in artillery studies it is agreed to use the quantity E_x for the estimation of all deviations. In the following, we shall find out why this usually does not lead to any difficulties in practice.

§ 25. Theorems on the standard deviation

We shall now show that the standard deviation actually possesses special properties which make it preferable to every other measure of the magnitude of dispersion—mean deviation, probable (i.e. equally likely) deviation, etc. As we shall realize a little later, the following problem is of fundamental value for applications.

Suppose that we have several random variables $x_1, x_2, \ldots, x_n$ with standard deviations $q_1, q_2, \ldots, q_n$. We set $x_1 + x_2 + \ldots + x_n = X$ and ask ourselves how to find the standard deviation Q of the quantity X if $q_1, q_2, \ldots, q_n$ are given and if we assume the random variables x_i $(1 \le i \le n)$ to be mutually independent.

By virtue of the theorem on the addition of mean values, we have

$$\bar{X} = \bar{x}_1 + \bar{x}_2 + \ldots + \bar{x}_n$$

and, consequently,

$$X - \bar{X} = (x_1 - \bar{x}_1) + (x_2 - \bar{x}_2) + \ldots + (x_n - \bar{x}_n),$$

from which it follows that

$$(X - \bar{X})^2 = \left[\sum_{i=1}^{n} (x_i - \bar{x}_i) \right]^2$$

$$= \sum_{i=1}^{n} (x_i - \bar{x}_i)^2 + \sum_{i=1}^{n} \sum_{\substack{k=1 \\ i \ne k}}^{n} (x_i - \bar{x}_i) \cdot (x_k - \bar{x}_k). \qquad (1)$$

We now note that

$$\overline{(X - \bar{X})^2} = Q^2, \qquad \overline{(x_i - \bar{x}_i)^2} = q_i^2 \quad (1 \le i \le n);$$

using the rule for the addition of mean values, we therefore find that

$$Q^2 = \sum_{i=1}^{n} q_i^2 + \sum_{i=1}^{n} \sum_{\substack{k=1 \\ i \ne k}}^{n} \overline{(x_i - \bar{x}_i) \cdot (x_k - \bar{x}_k)}. \qquad (2)$$

But, since the quantities x_i and x_k, according to our hypothesis are mutually independent for $i \ne k$, we have, by the rule for the multiplication of the mean values of mutually independent quantities,

$$\overline{(x_i - \bar{x}_i) \cdot (x_k - \bar{x}_k)} = \overline{(x_i - \bar{x}_i)} \cdot \overline{(x_k - \bar{x}_k)}$$

for $i \ne k$. Here, both factors in the right member are equal to zero inasmuch as, for instance,

$$\overline{(x_i - \bar{x}_i)} = \bar{x}_i - \bar{x}_i = 0;$$

thus, in the last sum in equality (2), each term separately vanishes, which leads us to the relation

$$Q^2 = \sum_{i=1}^{n} q_i^2,$$

i.e., *the variance of the sum of mutually independent random variables equals the sum of their variances.*

We see that in the case of mutually independent random variables, there is adjoined to the rule for the addition of mean values the very important *rule for the addition of variances*; for standard deviations, we obtain from this

$$Q = \sqrt{\sum_{i=1}^{n} q_i^2}. \tag{3}$$

This possibility simply expresses the standard deviation of a sum in terms of the standard deviations of its summands in the case in which they are mutually independent, and it represents one of the most important reasons for preferring the standard deviation in comparison with the mean, probable, or other deviations.

EXAMPLE 1.   For n shots with the same probability p of a hit, the mean value of the number of hits equals np (as we saw on page 77). In order initially to estimate how large the deviation of the actual number of hits from this mean value can turn out to be, we shall find the standard deviation of the number of hits; this is most simply done by applying formula (3).

In fact, the number of hits for n shots can be considered as the number of hits among n single shots (we already did this on page 76), and, since we consider these n outcomes to be mutually independent random variables, by the rule for the addition of variances we can make use of formula (3) to calculate the standard deviation Q of the total number of hits, where in (3) q_1, q_2, . . ., q_n denote the standard deviations of the number of hits for individual shots. But, the number of hits x_i for the ith shot is defined by the table

1	0
p	$1-p$

Therefore, $\bar{x}_i = p$ and $q_i^2 = \overline{(x - \bar{x}_i)^2} = (1-p)^2 p + p^2(1-p) = p(1-p)$; consequently,

$$Q = \sqrt{\sum_{i=1}^{n} q_i^2} = \sqrt{np(1-p)}.$$

This completes the solution of the problem posed.   Comparing the mean value np of the number of hits with its standard deviation $\sqrt{np(1-p)}$, we see that for large values of n (i.e., for a large number of shots) the latter is significantly less than the first and constitutes only a small portion of it.   Thus, for $n = 900$, $p = 1/2$, the mean value of the number of hits equals 450 and the standard deviation equals

$\sqrt{900 \cdot 1/2 \cdot 1/2} = 15$, so that the actual number of hits will deviate from its mean value by approximately only 3–4%.

EXAMPLE 2.   Let us imagine that a certain mechanism is being assembled which consists of n parts put together one right next to the other along a straight line and held together at the ends by some encompassing part (see Fig. 10).   The length of each part can differ somewhat from the corresponding standard and is therefore a random variable.   We assume these random variables to be independent.   If the average lengths of the parts and the standard deviations of these lengths are equal to

$$a_1, a_2, \ldots, a_n \quad \text{and} \quad q_1, q_2, \ldots, q_n,$$

respectively, then the mean value and standard deviation of the length of the chain consisting of n parts are equal respectively to

$$a = \sum_{k=1}^{n} a_k \quad \text{and} \quad q = \sqrt{\sum_{k=1}^{n} q_k^2}.$$

In particular, if $n=9$, $a_1=a_2=\ldots=a_9=10$ cm. and $q_1=q_2=\ldots=q_9=0.2$ cm., then $a=90$ cm. and $q=\sqrt{9 \cdot (0.2)^2}=0.6$ cm.

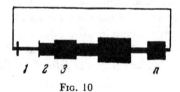

1   2   3   *n*

FIG. 10

We thus see that if on the average the length of every individual part deviates from its mean value by 2%, then the length of the chain consisting of these parts differs from its mean value by approximately $\frac{2}{3}$% only.   This situation—i.e., the decrease in the relative error upon addition of random variables—plays a significant role in the assembly of precision mechanisms.   In fact, if there were no mutual compensation of the deviations of the dimensions of the individual parts from the prescribed normal dimensions, then in the assembly of mechanisms one would continually encounter cases when the encompassing parts would not hold together the chain to be enveloped; or, conversely, in this connection, there would remain extraordinarily large gaps.   In both cases, an obvious reject would be obtained.   To combat this rejection by means of lessening the "tolerances" (i.e., by means of lessening the admissible deviations of the actual dimensions of the parts from those prescribed) would not be expedient, because a comparatively small increase in the precision of the parts would greatly increase their cost.[1]

EXAMPLE 3.   We assume that n measurements of a certain quantity are made under uniform conditions.   As a result of an entire series of circumstances (e.g., the position of the instrument, the observer, the state of the atmosphere, the presence in it of dust, and so forth), the different measurements will yield, generally speaking, distinct results—there are *random errors* in the measurement.   We shall denote the results of measurement by $x_1, x_2, \ldots, x_n$, assigning to each x the value of its measurement.   The mean value of all these random variables is the same, $\bar{x}$.   It is also natural, obviously, to assume the standard deviation q to be the same for all measurements, inasmuch as they are made under the same conditions.   Finally, we assume, as usual, that the quantities $x_1, x_2, \ldots, x_n$ are mutually independent.

We now consider the arithmetic mean

$$\xi = \frac{x_1 + x_2 + \ldots + x_n}{n}$$

of the results of n measurements.   This is a random variable; we shall find its mean value and standard deviation.   According to the addition rule,

$$\bar{\xi} = \frac{1}{n} \overline{(x_1 + x_2 + \ldots + x_n)} = \frac{1}{n} (\bar{x}_1 + \bar{x}_2 + \ldots + \bar{x}_n) = \frac{1}{n} (n\bar{x}) = \bar{x};$$

i.e., the mean value, as was essentially clear already earlier, is the same as for each individual measurement.

Further, the standard deviation of the sum $x_1 + x_2 + \ldots + x_n$ is equal, by the rule for the addition of variances (3), to

$$Q = \sqrt{nq^2} = q\sqrt{n},$$

and the standard deviation of the quantity ξ, comprising $(1/n)$th of this sum, equals

$$\frac{Q}{n} = \frac{q}{\sqrt{n}}.$$

We have arrived at an important result: *the arithmetic mean of* n *mutually independent and identically distributed random variables has*:

a) *the same mean value as each of the component quantities;*

b) *a standard deviation which is* $1/\sqrt{n}$ *times as large as each of the component standard deviations.*

[1] In recent years, technological thinking has come around to the conclusion that the creation of a theory of tolerances, based on argumentation and results from the theory of probability, is necessary.   This theory of tolerances is now being vigorously developed by Soviet scientists.

Thus, if the mean value of the quantity being measured is $\bar{x} = 200$ m. and the standard deviation is $q = 5$ m., then the arithmetic mean ξ of a hundred results of measurement will of course have the same number, 200 m., as its mean value; but its standard deviation will be $(1/\sqrt{100})$th $= (1/10)$th as large as for the individual measurement; i.e., it will amount to $q/10 = 0.5$ m. Thus, one has reason to expect that the arithmetic mean of a hundred actual results of measurement will be significantly closer to the mean value 200 m. than the result of any individual measurement. *The arithmetic mean of any number* n *of mutually independent quantities, each having the same variance* q², *possesses a variance equal to* q²/n.

CHAPTER 11

LAW OF LARGE NUMBERS

§ 26. Chebyshev's inequality

We have already asserted many times that the knowledge of any of the measures of dispersion of a random variable (for instance, its standard deviation) enables us to obtain a first approximation of how large the deviations (of the actual values of this quantity from its mean value) must be expected to be. However, this observation in itself is still devoid of any quantitative evaluation and does not enable us even to calculate approximately how probable large deviations can turn out to be. All this motivates us to carry out the following straightforward analysis, which was first carried out by Chebyshev. We start with the expression for the variance of a random variable x (see page 84):

$$Q_x^2 = \sum_{i=1}^{k} (x_i - \bar{x})^2 p_i.$$

Let α be an arbitrary positive number; if we discard all terms where $|x_i - \bar{x}| \leq \alpha$ in the above sum and retain only those for which $|x_i - \bar{x}| > \alpha$, this sum can only decrease as a result:

$$Q_x^2 \geq \sum_{|x_i - \bar{x}| > \alpha} (x_i - \bar{x})^2 p_i.$$

But this sum decreases still more if we replace the factor $(x_i - \bar{x})^2$ in each of its terms by the smaller quantity α^2:

$$Q_x^2 \geq \alpha^2 \sum_{|x_i - \bar{x}| > \alpha} p_i.$$

The sum now appearing in the right member is the sum of the probabilities of all those values x_i of the random variable x which deviate from $\bar{x}$ in one or the other direction by more than α; by the addition law, this is the probability that the quantity x take on any one of these values. In other words, this is the probability $P(|x - \bar{x}| > \alpha)$ that the deviation actually obtained be greater than α; we thus find

$$P(|x - \bar{x}| > \alpha) \leq \frac{Q_x^2}{\alpha^2}, \tag{1}$$

which enables us to estimate the probability of deviations greater than an arbitrary prescribed number α, provided the standard deviation Q_z is known. True—the estimate given by "Chebyshev's inequality" (1) frequently turns out to be very crude; nevertheless, it can be used directly in practice, not to mention that it is of very great theoretical significance.

At the end of the preceding section, we considered the following example: the mean value of the results of measurement is 200 m., the standard deviation is 5 m. Under these conditions, the probability of actually obtaining a deviation greater than 3 m. is noticeably high (one can imagine it to be greater than a half; its exact value can of course be found only when the distribution law of the results of the measurement is completely known). But we saw that for an arithmetic mean ξ of a hundred results of measurement, the standard deviation amounts in all to 0.5 m. Therefore, by virtue of inequality (1), we have

$$P(|\xi-200| > 3) \leq \frac{(0.5)^2}{3^2} = \frac{1}{36} \approx 0.03.$$

Thus, for the arithmetic mean of 100 measurements, the probability of obtaining a deviation of more than 3 m. is already very small. (In fact, it is even much smaller than the bound we obtained, so that in practice one can completely disregard the possibility of such a deviation.)

In Example 1 on page 89, we had a mean value of 450 and standard deviation 15 for the number of hits with 900 shots; for the probability that the actual number m of hits will be contained, for instance, between 400 and 500 (i.e., $|m-450| \leq 50$), the Chebyshev inequality yields

$$P(|m-450| \leq 50) = 1 - P(|m-450| > 50) \geq 1 - \frac{15^2}{50^2} = 0.91.$$

In fact, the actual probability is significantly larger than this.

§ 27. Law of large numbers

Suppose we have n mutually independent random variables $x_1, x_2, \ldots, x_n$ with the same mean value a and the same standard deviation q. For the arithmetic mean of these quantities,

$$\xi = \frac{x_1 + x_2 + \ldots + x_n}{n},$$

as we saw on page 91, the mean value is equal to a and the standard deviation equals $q/\sqrt{n}$; therefore, Chebyshev's inequality yields

$$P(|\xi - a| > \alpha) \leq \frac{q^2}{\alpha^2 n}. \tag{2}$$

for arbitrary positive α.

Suppose, for instance, that we are dealing with the arithmetic mean of n measurements of a certain quantity and suppose, as we had before, that $q = 5$ m., $a = 200$ m. We then obtain

$$P(|\xi - 200| > \alpha) \leq \frac{25}{\alpha^2 n}.$$

We can choose α very small—for example, $\alpha = 0.5$ m.; then

$$P(|\xi - 200| > 0.5) \leq \frac{100}{n}.$$

If the number n of measurements is very large, then the right-hand member of this inequality is arbitrarily small; thus, for $n = 10,000$, it equals 0.01 and we have

$$P(|\xi - 200| > 0.5) \leq 0.01$$

for the arithmetic mean of 10,000 measurements. If we agree to disregard the possibility of events having such small probabilities, then we can assert that for 10,000 measurements, their arithmetic mean will certainly be different from 200 m. in one or the other direction by not more than 50 cm. If we wished to attain a still better approximation, for instance to 10 cm., then it would be necessary to set $\alpha = 0.1$ m. and we would obtain

$$P(|\xi - 200| > 0.1) \leq \frac{25}{0.01 n} = \frac{2500}{n}.$$

In order to make the right-hand member of this inequality less than 0.01, we would have to take a number of measurements equal not to 10,000 (which is now insufficient), but rather to 250,000. It is clear that in general we can, no matter how small α is, make the right-hand member of inequality (2) as small as we please—to this end, we must only take n sufficiently large. Thus, for sufficiently large n we can assume that the reverse inequality

$$|\xi - a| \leq \alpha$$

is arbitrarily close to certainty.

If the random variables $x_1, x_2, \ldots, x_n$ *are mutually independent and if they all have the same mean value* a *and the same standard deviation, then the quantity*

$$\xi = \frac{x_1 + x_2 + \ldots + x_n}{n}$$

will differ from a *by an arbitrarily small amount for sufficiently large* n *with a probability which is as close to unity as we please (i.e., practically certain).*

This is the simplest particular case of one of the most fundamental theorems of probability theory—the so-called *law of large numbers,* which was discovered in the middle of the last century by the great Russian mathematician Chebyshev. The profound content of this remarkable law consists in that, whereas an individual random variable can (as we know) frequently take on values which are far removed from its mean value (i.e., it can have significant dispersion), the arithmetic mean of a large number of random variables behaves in this relation completely differently: such a quantity has very little dispersion—it assumes, with very high probability, only values which are close to its mean value. This of course occurs because upon taking the arithmetic mean the random deviations in one or the other direction mutually cancel each other out, as a result of which the total deviation turns out to be small *in the majority of cases.*

The important and frequently encountered application of the results of the Chebyshev theorem which we just proved consists in being able to judge, after a comparatively small test (or sample), the quality of a large quantity of homogeneous material. Thus, for example, the quality of cotton found in a bale is judged by several small bunches (i.e., samples) pulled out at random from various places in the bale. Or, the quality of a large lot of grain is judged by several small scales (measures) filled with grains caught at random by the scales from various spots of the lot being evaluated.[1] Judgments of the quality of production, made on the basis of such a choice, possess great accuracy since the number of grains caught in the scales, say, although small in comparison with the entire supply of grain, is in itself large and enables one, according to the law of large numbers, to judge sufficiently accurately the mean weight of one grain and hence the quality of the entire lot of grain. In exactly the same way, one judges a twenty-*pood*[2] bale of cotton by a small sample containing several hundred fibers which weigh, altogether, some decimal part of a gram.

---

[1] The scales catch, say, 100–200 grains and the entire lot contains tens and perhaps even hundreds of tons of grain.

[2] One *pood* equals approximately 16.38 kg.

§ 28. Proof of the law of large numbers

Up to this point, we considered only the case in which all the quantities $x_1, x_2, \ldots$ have the same mean value and the same standard deviation. The law of large numbers, however, is applicable under much more general assumptions. We shall now consider the case when the mean values of the quantities $x_1, x_2, \ldots$ can be arbitrary numbers (we shall denote them by $a_1, a_2, \ldots$, respectively), which, generally speaking, may be distinct. Then the mean value of the quantity

$$\xi = \frac{1}{n} (x_1 + x_2 + \ldots + x_n)$$

will obviously be the quantity

$$A = \frac{1}{n} (a_1 + a_2 + \ldots + a_n),$$

where, by virtue of Chebyshev's inequality (1),

$$P(|\xi - A| > \alpha) \leq \frac{Q_\xi^2}{\alpha^2} \tag{3}$$

for arbitrary positive α. We see that everything reduces to estimating the quantity Q_ξ^2—but here it is almost as simple to estimate this quantity as in the particular case considered earlier. Q_ξ^2 is the variance of the quantity ξ which is equal to the sum of n mutually independent random variables divided by n (we of course retain the assumption of mutual independence). By the law of addition of variances, we therefore have

$$Q_\xi^2 = \frac{1}{n^2} (q_1^2 + q_2^2 + \ldots + q_n^2),$$

where $q_1, q_2, \ldots$ denote the standard deviations of the quantities $x_1, x_2, \ldots$, respectively. We shall now also assume that these standard deviations may be, generally speaking, distinct. However, we shall admit nonetheless that no matter now many quantities we take (i.e., no matter how large the number n is), the standard deviations of these random variables will all be less than some positive number b. In practice, this condition turns out always to be satisfied, since one must combine quantities of more or less the same type and the degree of their dispersion turns out to be not too different for distinct quantities. Thus, we shall assume that $q_i < b$ $(i = 1, 2, \ldots)$; but then the last equality gives us

$$Q_\xi^2 < \frac{1}{n^2} nb^2 = \frac{b^2}{n},$$

as a consequence of which we conclude from inequality (3) that

$$P(|\xi - A| > \alpha) < \frac{b^2}{n\alpha^2}.$$

No matter how small α might be, for a sufficiently large number n of random variables taken, the right-hand member of this inequality can be made arbitrarily small; this, evidently, proves the law of large numbers in the general case which we just considered.

Thus, *if the quantities* x_1, x_2, ... *are mutually independent and all their standard deviations remain less than the same positive number, then, for sufficiently large* n, *one can expect arbitrarily small (in absolute value) deviations for the arithmetic mean*

$$\xi = \frac{1}{n}(x_1 + x_2 + \ldots + x_n)$$

with a probability as close to unity as we please.

This is the law of large numbers in the general form given it by Chebyshev.

It is now appropriate to turn our attention to one essential situation. We assume that a certain quantity a is being measured.   Repeating the measuring, under the same conditions, the observer obtains the numerical results x_1, x_2, ..., x_n which do not coincide completely. As an approximate value of a, he takes the arithmetic mean

$$a \approx \frac{1}{n}(x_1 + x_2 + \ldots + x_n).$$

One asks, can one hope to obtain a value of a which is as precise as we please by carrying out a sufficiently large number of trials?

This will be the case if the measurements are carried out without a systematical error; i.e., if

$$\bar{x}_k = a \quad (\text{for } k = 1, 2, \ldots, n)$$

and if the values themselves do not possess any indeterminacy—in other words, if in making the measurements, we read those indications on the instrument which are in reality obtained there.   But if the instrument is constructed so that it cannot yield readings more accurate than a certain quantity δ, for instance because of the fact that the width of the scale division on which the reading is made equals δ, then it is understood that one cannot hope to obtain accuracy greater than $\pm \delta$.   It is clear that in this case the arithmetic mean will also possess the same indeterminacy δ as does each of the x_k's.

The remark just made teaches us that if instruments give us results of measurements with a certain indeterminacy δ, then to strive by means of the law of large numbers to obtain a value *a* with greater accuracy is a delusion and the calculations themselves made under these circumstances become an empty arithmetic exercise.

CHAPTER 12

NORMAL LAWS

§ 29. Formulation of the problem

We saw that a significant number of natural phenomena, production processes, and military operations occur under the essential influence of certain random variables. Frequently, owing to the fact that a phenomenon, process, or operation is not determinate, all that we may know about these random variables is their laws of distribution; i.e., the itemization of their possible values with an indication of the probability of each of these values. If the quantity can be assigned an infinite set of distinct values (the range of a shell, the size of error in a measurement, and so forth), then it is preferable to indicate the probability not of the individual values of it, but of entire portions of such values (for example, the probability that an error of measurement lie between − 1 mm. and + 1 mm., between 0.1 mm. and 0.25 mm., and so on.) These considerations do not modify the essence of the matter at hand—in order to make the most effective use of random variables, we must obtain the most precise presentation possible of their laws of distribution.

If, endeavoring to learn the distribution laws of the random variables which we encounter, we reject all discussions and conjectures of a general nature, approach every random variable without any preliminary suppositions, and strive to find by purely experimental means all features of the distribution law peculiar to it, then we would set before ourselves a problem which is almost impossible to solve without a great expenditure of labor. In every new case, a large number of trials would be required in order to establish even the more important features of a new, unknown distribution law. Therefore, scientists have for a long time now been striving to find such general types of distribution laws whose presence could reasonably be predicted, expected, surmised, if not for all, then at least for extensive classes of random variables encountered in practice. A long time ago, just such types were established theoretically and their existences were subsequently confirmed experimentally. It is readily understood how convenient it is to be able on the basis of theoretical considerations to predict in advance of what type must be the distribution law for a random variable that we may encounter. If such a conjecture turns out to be correct, then usually a very small number of trials or observations is sufficient to establish all necessary features of the distribution law sought.

Theoretical investigations have shown that in a large number of cases encountered in practice we can with sufficient justification anticipate distribution laws of a definite type. These laws are called *normal laws*. We shall briefly discuss these laws in the present chapter —omitting all proofs and precise formulations because of their difficulty.

Among the random variables which we encounter in practice, very many have the nature of "random discrepancies" or of "random errors" or at least they can be easily reduced to such "errors." Suppose, for example, that we are studying the range x of a shell fired from some cannon. We naturally assume that there exists some average range x_0 on which we set the aiming instruments; the difference $x - x_0$ constitutes the "error" or "discrepancy" in the range, and the study of the random variable x reduces entirely and directly to the investigation of the "stochastic error" $x - x_0$. But such an error, whose magnitude varies from shot to shot, depends as a rule on very many causes which act independently of one another: the random vibrations of the barrel of the cannon, the unavoidable (even though small) difference in the weight and form of the shells, the random variation of atmospheric conditions causing variations in air resistance, the random errors in aiming (if aiming is made anew before every firing or before every short sequence of firings)—all these and many other causes are capable of producing errors in the range. All these individual errors will be mutually independent random variables, in which connection they are such that *the action of each of them constitutes only a very small portion of their collective action* and the final error $x - x_0$, which we wish to investigate, will simply be the total effect of all these stochastic errors resulting from the individual causes. Thus, in our example the error of interest to us is the sum of a large number of independent random variables and it is clear that the situation will be similar for the majority of stochastic errors with which we deal in practice.

The theoretical discussion, which we cannot reproduce here, shows that the distribution law of a random variable which is the sum of a very large number of mutually independent random variables must be close to a law of some definite type—indeed, by virtue of the fact alone that, whatever the nature of the terms, *each of them is small in comparison with the entire sum.*[1] And this type is the type of normal laws.

It is thus possible for us to assume that a very significant portion of the random variables which we encounter in practice (i.e., all errors composed of a large number of mutually independent stochastic errors) is approximately distributed according to normal laws. We must now become acquainted with the fundamental features of these laws.

§ 30. Concept of a distribution curve

In § 15 we already had the opportunity to represent distribution laws graphically—with the aid of a diagram; this method is very useful inasmuch as it enables us at a glance, without recourse to the study of

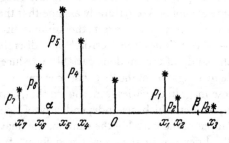

Fig. 11

tables, to grasp the important features of the distribution law being investigated. The scheme of this representation is as follows: on a horizontal straight line we mark off the different possible values of the given random variable, starting with some reference point 0— positive values are marked off to the right and negative ones to the left (see Fig. 11). At each such possible value we plot along a vertical, upward, the probability of this value. The scale in both directions is chosen so that the entire diagram has a convenient and easily readable form. A quick glance at Fig. 11 convinces us of the fact that the random variable which it characterizes has the most probable value x_5 (which is negative) and that as the possible values of this quantity depart from the number x_5 their probabilities continually (and rather rapidly) decrease. The probability that the quantity take on a value included in some interval (α, β) is equal, according to the addition rule, to the sum of the probabilities of all possible values lying in this

---

[1] In this connection, see also the Conclusion.

segment and is geometrically represented by the sum of the lengths of the vertical lines situated on this segment; in Fig. 11, $P(\alpha < x < \beta)$ $= p_1 + p_2 + p_4 + p_5$. If the number of possible values is very large, as frequently happens in practice, then, in order that the line not extend too far along the horizontal, we take a very small scale in the horizontal direction, as a consequence of which the possible values are arranged extremely densely (Fig. 12) so that the tips of the vertical lines drawn merge, as far as our eye is concerned, into one dense curve, which is called the *distribution curve* of the given random variable. And here, of course, the probability of the inequality $\alpha < x < \beta$ is represented graphically as the sum of the lengths of the vertical lines situated on the segment (α, β). We now assume that the distance between two neighboring possible values is always equal to unity;

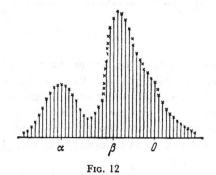

FIG. 12

this will be the case, for instance, if the possible values are expressed by a sequence of successive integers, which in practice can always be attained by choosing a sufficiently fine unit in our scale. Then the length of every vertical line is numerically equal to the *area of the rectangle* for which this line serves as the height and whose base equals its unit distance from the neighboring line (Fig. 13). Thus, the probability of the inequalities $\alpha < x < \beta$ can be expressed graphically as the sum of the areas of rectangles, drawn in the figure, situated on these segments. But if the possible values are arranged very densely as in the preceding Fig. 12, then the sum of the areas of such rectangles does not differ practically from the areas of the curvilinear figures bounded below by the segment (α, β), above by the distribution curve, and on the sides by the vertical lines drawn at the points α and β (see Fig. 14).[1] Thus, on a curvilinear diagram of the type shown in

---

[1] In this connection, of course, we take, as before, as unity the length of the distance between two neighboring possible values.

Fig. 14, the probability that the given random variable fall into some
segment is simply and conveniently expressed by the area lying over
this segment and below the distribution curve. If the distribution
law of the given quantity is given by such a curvilinear diagram, then
we do not draw vertical lines on it which are of no value and would
only obstruct the figure. And even the question of the probabilities
of individual values would in this situation lose its reality; if there are

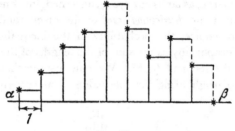

FIG. 13

very many possible values (this lies at the basis of all curvilinear
diagrams), then the probabilities of individual values will be, as a rule,
negligibly small (practically equal to zero) and are devoid of all
interest. Thus, in measuring the distance between inhabited points,
it is not at all important to know that the result of measurement
deviates from the true value by exactly 473 cm. In contrast, the

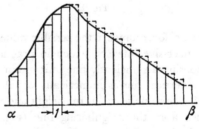

FIG. 14

question of the probability that the deviation is contained in the
interval from 3 m. to 5 m. is of essential interest. And so in all such
cases we conclude that if the random variable takes on very many
values, then it is important for us to know the probability not of the
individual values but the probability of entire segments of such values.
But it is namely these latter probabilities which can be pictorially and
directly represented by areas on curvilinear diagrams, as we have just
seen.

§ 31. Properties of normal distribution curves

A quantity which is distributed according to a normal law always has an infinite set of possible values; therefore, it is convenient to express the normal laws by means of curvilinear diagrams. In Fig. 15, there are shown several distribution curves for normal laws. Disregarding all differences in the appearance of these curves, we see in them the clearly expressed features common to all of them:

1) All the curves have one highest point such that if we depart from it to the right or left they continually decrease. Clearly, this means that upon departure of the values of the random variable from its most probable value, their probabilities continually decrease.

2) All the curves are symmetrical with respect to a vertical line drawn through the highest point. Clearly, this means that the values which are equally distant from the most probable value have the same probability.

3) All the curves have a bell-shaped form: in the vicinity of the highest point they are concave downward and at some distance from the highest point they bend and become concave upward. This distance, as well as the maximal height, is different for distinct curves.[1]

In what respect do distinct normal curves differ from one another? In order to answer this question clearly, we must first of all recall that for every distribution curve all the area situated under it equals unity, because this area equals the probability that the given random variable take on some one of its values; i.e., the probability of a certain event. The difference of individual distribution curves from one another therefore consists only in that this total area, which is the same for all curves, is distributed differently over various parts of the curve. For normal laws, as the curves in Fig. 15 show, the question basically consists of ascertaining what portion of this total area is concentrated on the parts which are immediately adjacent to the most probable value and what portion on parts at a greater distance from this value. For the law represented in Fig. 15(a), almost all the area is concentrated

---

[1] For readers who are familiar with the elements of higher mathematics, we note that the equation of the curve, expressing a normal law, has the form

$$y = \{1/(\sigma\sqrt{2\pi})\} \cdot \exp\{-(x-a)^2/(2\sigma^2)\},$$

where exp α means the number $e = 2.71828\ldots$ —the base of natural logarithms— to the power α; $\pi = 3.14159\ldots$ is the ratio of the length of a circumference of a circle to its diameter; and the quantities a, σ^2 are the mean value and variance of the random variable. Knowledge of the analytic form of the normal law can make it much easier for the reader to become acquainted with the further material in the book. However, the discussion of all that follows is also accessible to the reader who is not at all acquainted with higher mathematics.

in the immediate vicinity of the most probable value; this means that
the random variable with predominant probability—and, hence,
in the overwhelming majority of cases—takes on values near its most
probable value. Because, by virtue of the symmetry mentioned
above, in the case of a normal law the most probable value always
coincides with the average value, we can say that the random variable
subject to law (a) is not dispersed much; in particular, its variance and
standard deviation are small.

Conversely, in the case shown in Fig. 15(c), the area concentrated
in the immediate vicinity of the most probable value comprises only a
small portion of the total area (we see at once the difference if we

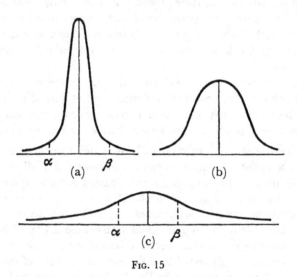

Fig. 15

juxtapose on Fig. 15(a) and Fig. 15(c) the portions (α, β) of the same
length and the areas situated on them). Here, it is very probable,
therefore, that the random variable will take on values which deviate
significantly from its most probable value. The quantity is extremely
dispersed; its variance and standard deviation are large.

Case (b), obviously, occupies a position intermediate between cases
(a) and (c).

In order to become acquainted in the quickest way with the entire
set of normal laws and to learn how to apply these laws, it will be
expedient for us to proceed from two fundamental properties of these
laws. These properties, which will now be formulated in detail,
cannot be proved here since to do this we would have first of all to

define the normal laws precisely, which would require of the reader knowledge of higher mathematics.

PROPERTY 1.   If the quantity x is distributed according to a normal law, then,

1) for arbitrary constants $c > 0$ and d, the quantity $cx + d$ is also distributed according to some normal law; and,

2) conversely, given an arbitrary normal law, a (unique) pair of numbers $c > 0$ and d can be found such that the quantity $cx + d$ is distributed precisely according to this law.

Thus, if the random variable x is distributed according to a normal law, then the distribution laws to which the quantities $cx + d$ are subject, for all possible values of the constants $c > 0$ and d, comprise all normal laws.

PROPERTY 2.   If the random variables x and y are mutually independent and are distributed according to normal laws, then their sum $z = x + y$ is also distributed according to some normal law.

Assuming these two fundamental properties without proof, we can now rigorously establish a number of properties of normal laws which follow from them—these properties are especially important in practice.

I. *For any two numbers* a *and* q > 0 *there exists a unique normal law with mean value* a *and standard deviation* q.

In fact, suppose x is a random variable, distributed according to a normal law with mean value $\bar{x}$ and standard deviation Q_x. On the basis of Property 1, our assertion will be proved if we show that there exists a unique pair of numbers $c > 0$ and d satisfying the requirement that the quantity $cx + d$ has the mean value a and standard deviation q.   If the table of values of x has the form

TABLE I

x_1	x_2	...	x_n
p_1	p_2	...	p_n

then to the quantity $cx + d$ (where $c > 0$ and d are, for the time being, arbitrary numbers) there will correspond the table

$cx_1 + d$	$cx_2 + d$	...	$cx_n + d$
p_1	p_2	...	p_n

Clearly, $\sum\limits_{k=1}^{n} x_k p_k = \bar{x}$, $\sum\limits_{k=1}^{n} (x_k - \bar{x})^2 p_k = Q_x^2$. Our requirements reduce to the two requirements:

$$\sum_{k=1}^{n} (cx_k + d) p_k = a; \qquad \sum_{k=1}^{n} (cx_k + d - a)^2 p_k = q^2.$$

The first of these conditions yields

$$c \sum_{k=1}^{n} x_k p_k + d \sum_{k=1}^{n} p_k = a,$$

or,

$$c\bar{x} + d = a, \tag{1}$$

and the second yields

$$\sum_{k=1}^{n} (cx_k + d - c\bar{x} - d)^2 p_k = c^2 \sum_{k=1}^{n} (x_k - \bar{x})^2 p_k = c^2 Q_x^2 = q^2,$$

from which it follows (since c is > 0) that

$$c = q/Q_x, \tag{2}$$

and, hence, (1) implies that

$$d = a - c\bar{x} = a - q\bar{x}/Q_x. \tag{3}$$

Thus, for given a and q, the numbers c and d can always be found with the aid of formulas (2) and (3) and they are unique; also, the quantity $cx + d$ is subject to the normal law with mean value a and standard deviation q; with this, our assertion is completely proved.

If we do not restrict ourselves to normal laws, but rather consider all possible distribution laws, then the prescription of the mean value and variance or the standard deviation of a random variable still gives us very little information about its distribution law since there exist very many (and in this connection essentially distinct) distribution laws which possess the same mean value and the same variance; in the general case, the prescription of the mean value and variance only very approximately characterizes for us the distribution law of the given random variable.

The situation is quite different if we agree to restrict ourselves to the consideration of only normal laws. On the one hand, as we have just convinced ourselves, an arbitrary assumption concerning the mean value and variance of a given random variable is compatible with the requirement that it be subject to a normal law. On the other hand— and this is most important—if we have reason to assume in advance that the given quantity is distributed according to one of the normal laws, then the prescription of its mean value and variance uniquely determines its distribution law so that its nature as a random variable becomes completely known.   In particular, knowing the mean value and the variance of such a quantity, we can calculate the probability that its value will belong to some arbitrarily chosen interval.

II. *The ratio of probable deviation to the standard deviation is the same for all normal laws.*

Suppose we have two arbitrary normal laws and let x be a random variable subject to the first of these laws.   In virtue of the fundamental Property 1, there exist constants $c > 0$ and d such that the quantity $cx + d$ is distributed in accordance with the second of the given laws. We denote the standard deviation and the probable deviation of the first quantity by Q_x and E_x, respectively, and the same deviations of the second quantity by q and e, respectively.   By the definition of the probable deviation, we have

$$P(|(cx+d)-(c\bar{x}+d)| < e) = 1/2,$$

or

$$P(c|x-\bar{x}| < e) = 1/2,$$

or, finally,

$$P\left(|x-\bar{x}| < \frac{e}{c}\right) = 1/2.$$

From this, again by virtue of the definition of the probable deviation, it follows that e/c is the probable deviation of the quantity x, i.e.,

$$\frac{e}{c} = E_x,$$

from which it follows that

$$\frac{e}{E_x} = c;$$

therefore, ratio (2) above shows that

$$\frac{e}{E_x} = \frac{q}{Q_x},$$

which implies that

$$\frac{e}{q} = \frac{E_x}{Q_x};$$

i.e., the ratio of the probable deviation to the standard deviation is the same for the two laws.   Since, by hypothesis, these laws were two arbitrary normal laws, our assertion is proved.

The ratio e/q is thus an absolute constant; we denote it by the letter λ; λ has been computed and found to be

$$\lambda = \sqrt{\frac{2}{\pi}} \approx 0.674.$$

This means that for an arbitrary normal law, we have

$$e = \sqrt{\frac{2}{\pi}}\, q.$$

By virtue of this exceptionally simple connection between the numbers e and q, it is immaterial in practice, for quantities distributed according to normal laws, which of the two distribution characteristics we shall use. We saw above that, generally speaking (i.e., even if we do not restrict ourselves to quantities which are distributed according to normal laws), the standard deviation possesses a whole string of simple properties which other measures of dispersion lack and which in the majority of cases force theoreticians as well as practitioners to choose precisely the standard deviation as the measure of dispersion. We also pointed out above that artillery personnel, nevertheless, almost always make use of mean deviations. We shall now see why this tradition cannot cause any serious consequences: those random variables with which artillery science and practice deal almost always turn out to be distributed according to normal laws, and, for such quantities, by virtue of the proportionality mentioned above, the choice of one or another characterization is immaterial in practice.

III. *Suppose* x *and* y *are mutually independent random variables, subject to normal laws, and that* z = x + y. *Then*

$$E_z = \sqrt{E_x^2 + E_y^2},$$

where E_x, E_y, E_z *denote the probable deviations of the quantities* x, y, z, *respectively.*

An analogous formula for standard deviations, as we know from § 25, holds, whatever may be the distribution laws of the quantities x and y. In the case in which these are normal laws, the quantity z, by virtue of the fundamental Property 2, is also distributed according to a normal law; therefore, in view of the preceding Property II, we have

$$E_x = \lambda Q_x, \qquad E_y = \lambda Q_y, \qquad E_z = \lambda Q_z,$$

and, hence,

$$E_z = \lambda\sqrt{Q_x^2 + Q_y^2} = \sqrt{(\lambda Q_x)^2 + (\lambda Q_y)^2} = \sqrt{E_x^2 + E_y^2}.$$

We see that in the case of normal laws, one of the most important properties of standard deviations carries over directly to probable (i.e., equally likely) deviations also.

§ 32. Solution of problems

We agree to call a law for which the mean value equals zero and whose variance equals unity a *fundamental normal law*. If x is a random variable possessing a fundamental normal law, then we agree, for the sake of brevity, to write

$$P(|x| < a) = \Phi(a)$$

for an arbitrary positive number a. Thus, $\Phi(a)$ is the probability that the quantity x which is subject to a fundamental normal law does not exceed in absolute value the number a. A very precise table has been constructed for the values of $\Phi(a)$, giving its values for various values of the number a. Such a table serves as an indispensable tool for everyone who has to deal with probability calculations. It is appended to every book devoted to probability theory. At the end of the present book (on page 123), the reader will also find such a table. Having a table of values of the function $\Phi(a)$ at hand, one can easily and with a great deal of precision carry out all calculations for arbitrary quantities distributed according to normal laws. We shall now show by means of examples how this is done.

PROBLEM 1. A random variable x is distributed according to a normal law with mean value $\bar{x}$ and with standard deviation Q_x. Find the probability that the deviation $x - \bar{x}$ does not exceed in absolute value the number a.

Let z be a random variable distributed according to a fundamental normal law. By virtue of the fundamental Property I (page 107), numbers $c > 0$ and d can be found such that the quantity $cz + d$ has mean value $\bar{x}$ and standard deviation Q_x; i.e., it is subject to the same normal law as the given quantity x. Therefore,

$$P(|x - \bar{x}| < a) = P(|(cz + d) - (c\bar{z} + d)| < a)$$
$$= P(c|z - \bar{z}| < a);$$

but, by virtue of formula (2) on page 108, we here have $c = Q_x/Q_z = Q_x$, since $Q_z = 1$ (the variance equals unity for a fundamental normal law). We therefore find that

$$P(|x - \bar{x}| < a) = P(Q_x|z - \bar{z}| < a)$$
$$= P(|z| < a/Q_x) = \Phi(a/Q_x). \tag{4}$$

This solves our problem, since the quantity $\Phi(a/Q_x)$ is found directly from the table. Thus, our table—with the aid of formula (4)—enables us to calculate easily the probability of any boundary of the deviation for a quantity which is subject to an arbitrary normal law.

EXAMPLE 1. Some part of a mechanism is prepared on a machine. It turns out that its length x represents a random variable distributed according to a normal law and it has mean value 20 cm. and standard deviation 0.2 cm. Find the probability that the length of the part lies between 19.7 cm. and 20.3 cm.; i.e., that the deviation in either direction does not exceed 0.3 cm.

By virtue of formula (4) and our table, we have

$$P(|x-20| < 0.3) = \Phi\left(\frac{0.3}{0.2}\right) = \Phi(1.5) = 0.866.$$

Thus, about 87% of all articles prepared under the given conditions will have lengths between 19.7 cm. and 20.3 cm.; the remaining 13% will have greater deviations from the average.

EXAMPLE 2. Under the conditions of Problem 1, find out to what precision the length of an article can be guaranteed with a probability of 0.95.

The problem consists, obviously, in finding a positive number a for which

$$P(|x-20| < a) > 0.95.$$

The calculations of Example 1 show that $a=0.3$ is too small here because in this case the left-hand member of the last inequality is less than 0.87. Since, according to equation (4),

$$P(|x-20| < a) = \Phi\left(\frac{a}{0.2}\right) = \Phi(5a),$$

we must first find in our table a value $5a$ for which

$$\Phi(5a) > 0.95.$$

We find that this will be the case for

$$5a > 1.97,$$

from which it follows that $a > 0.394$. Thus, we can guarantee, with a probability exceeding 0.95, that the deviation of the length will not exceed 0.4 cm.

EXAMPLE 3. In certain practical problems, we assume that the random variable x which is distributed according to a normal law does not reveal a deviation greater than three standard deviations Q_x. What basis do we have for this assertion?

Formula (4) and our table show that

$$P(|x-\bar{x}| < 3Q_x) = \Phi(3) > 0.997,$$

and, consequently, that

$$P(|x-\bar{x}| > 3Q_x) < 0.003.$$

In practice, this means that the deviations which surpass in absolute value $3Q_x$ will be encountered more rarely on the average than three times in a thousand. Is it possible to neglect such a probability or must it nevertheless be considered? This, of course, depends on the content of the problem and cannot be prescribed for all situations.

We note that the relation $P(|x-\bar{x}| < 3Q_x) = \Phi(3)$ is evidently a particular case of the formula

$$P(|x-\bar{x}| < aQ_x) = \Phi(a), \tag{5}$$

which follows from formula (4) and holds for every random variable x which is distributed according to a normal law.

EXAMPLE 4.   It is found in connection with the average weight of an article of 8.4 kg. that the deviations which in absolute value exceed 50 g. are encountered on the average three times in every 100 articles. Allowing that the weight of the articles is distributed according to a normal law, determine its probable deviation.

We are given

$$P(|x-8.4| > 0.05) = 0.03,$$

where x is the weight of an article chosen at random. It follows from this that

$$0.97 = P(|x-8.4| < 0.05) = \Phi\!\left(\frac{0.05}{Q_x}\right);$$

our table shows that $\Phi(a) = 0.97$ for $a \approx 2.12$. Therefore

$$\frac{0.05}{Q_x} \approx 2.12,$$

which implies that

$$Q_x \approx \frac{0.05}{2.12}.$$

The probable deviation, as we know, equals

$$E_x = 0.674 Q_x \approx 0.0155 \text{ kg.} = 15.6 \text{ g.}$$

EXAMPLE 5.   In firing from a certain cannon, the deviation of the shell from the target is due to three mutually independent causes: 1) the error in determination of the position of the target, 2) the error in aiming, and 3) the error from causes which vary from one shot to another (e.g., the weight of the shell, atmospheric conditions, and so forth). Assuming that all three errors are distributed according to normal laws with mean value 0 and that their probable deviations are equal to 24 m., 8 m., and 12 m., respectively, find the probability that the total deviation from the target does not exceed 40 m.

Since the probable deviation of the total error x is, by virtue of Property II (see page 109), equal to

$$\sqrt{24^2+8^2+12^2} = 28 \text{ m.,}$$

the standard deviation of the total error equals

$$\frac{28}{0.674} \approx 41.5$$

and, hence,

$$P(|x| < 40) = \Phi\left(\frac{40}{41.5}\right) \approx \Phi(0.964) = 0.665.$$

Deviations which do not exceed 40 m. will thus be observed in approximately 2/3 of all cases.

PROBLEM II. The random variable x is distributed according to the normal law with mean value $\bar{x}$ and standard deviation Q_x. Find the probability that the deviation $x-\bar{x}$ is in absolute value included between the numbers a and b $(0 < a < b)$.

Since, by the addition rule,

$$P(|x-\bar{x}| < b) = P(|x-\bar{x}| < a) + P(a < |x-\bar{x}| < b),$$

we have

$$P(a < |x-\bar{x}| < b) = P(|x-\bar{x}| < b) - P(|x-\bar{x}| < a)$$

$$= \Phi\left(\frac{b}{Q_x}\right) - \Phi\left(\frac{a}{Q_x}\right), \tag{6}$$

and this solves the problem posed.

In the great majority of problems in practice, this table of values of the quantity $\Phi(a)$ which we have been using all along proves to be, however, an unduly cumbersome calculation tool. It frequently turns out to be necessary to consider only the probability that the deviation $x-\bar{x}$ falls into more or less large intervals; therefore, it is desirable, for practical purposes, to have, alongside our "complete" table, also abridged tables which are easily constructed from the complete table with the aid of formula (6).

We now give an example of constructing this sort of table which is a great deal cruder than the table at the end of this book (p. 123), but which nevertheless is entirely sufficient in many cases.   We subdivide the entire interval of variation of the quantity $|x-\bar{x}|$ into five sub-intervals: 1) from 0 to $0.32Q_x$, 2) from $0.32Q_x$ to $0.69Q_x$, 3) from $0.69Q_x$ to $1.15Q_x$, 4) from $1.15Q_x$ to $2.58Q_x$, and 5) from $2.58Q_x$ to max $|x-\bar{x}|$.

Making use of formula (4), we find that

$$P(|x-\bar{x}| < 0.32Q_x) = \Phi(0.32) \approx 0.25$$
$$P(0.32Q_x < |x-\bar{x}| < 0.69Q_x) = \Phi(0.69) - \Phi(0.32) \approx 0.25$$
$$P(0.69Q_x < |x-\bar{x}| < 1.15Q_x) = \Phi(1.15) - \Phi(0.69) \approx 0.25$$
$$P(1.15Q_x < |x-\bar{x}| < 2.58Q_x) = \Phi(2.58) - \Phi(1.15) \approx 0.24$$
$$P(|x-\bar{x}| > 2.58Q_x) = 1 - \Phi(2.58) \approx 0.01.$$

It is convenient to depict the result of these calculations with the aid of a graphical scheme (see Fig. 16).   Here, the entire real line is

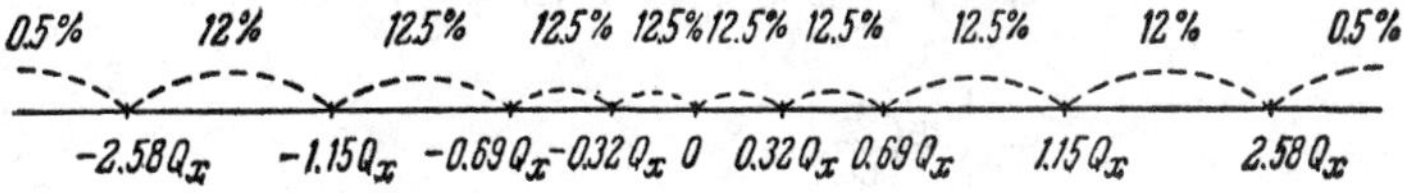

FIG. 16

subdivided into ten subintervals—five of which are positive and five negative.   Above each subinterval we have indicated what percentage of the deviations actually observed will on the average fall into this subinterval.   Thus, for example, according to the calculations made above, approximately 25% of all deviations should fall into the sub-intervals $(-1.15Q_x, -0.69Q_x)$ and $(0.69Q_x, 1.15Q_x)$ taken together. By virtue of the symmetry of the normal laws, the deviations will fall with approximately the same frequency into both of these sub-intervals, so that about 12.5% of the total number of deviations will fall into each of them.   Having at hand this simple scheme, or one similar to it, we can immediately visualize in a rough way the dis-tribution of the deviations for a random variable which is subject to a normal law with arbitrary mean value and standard deviation.

Finally, we consider how to calculate the probability that a random variable which is subject to a normal law lies in some arbitrarily prescribed subinterval.

PROBLEM III.   Knowing that the random variable x is distributed according to a normal law with mean value $\bar{x}$ and standard deviation

Q_x, calculate with the aid of the table the probability of the inequality $a < x < b$, where a and b $(a < b)$ are arbitrary, prescribed numbers.

We will have to consider three cases, depending on the disposition of the numbers a and b with respect to $\bar{x}$.

First case: $\bar{x} \leq a \leq b$.

According to the addition rule, we have

$$P(\bar{x} < x < b) = P(\bar{x} < x < a) + P(a < x < b),$$

from which it follows that

$$\begin{aligned} P(a < x < b) &= P(\bar{x} < x < b) - P(\bar{x} < x < a) \\ &= P(0 < x - \bar{x} < b - \bar{x}) - P(0 < x - \bar{x} < a - \bar{x}). \end{aligned}$$

But, by virtue of the symmetry of the normal laws, we have, for arbitrary $\alpha > 0$

$$\begin{aligned} P(0 < x - \bar{x} < \alpha) &= P(-\alpha < x - \bar{x} < 0) \\ &= \frac{1}{2} P(-\alpha < x - \bar{x} < \alpha) = \frac{1}{2} P(|x - \bar{x}| < \alpha) \\ &= \frac{1}{2} \Phi\left(\frac{\alpha}{Q_x}\right); \end{aligned} \tag{7}$$

therefore,

$$P(a < x < b) = \frac{1}{2}\left\{\Phi\left(\frac{b - \bar{x}}{Q_x}\right) - \Phi\left(\frac{a - \bar{x}}{Q_x}\right)\right\}.$$

Second case: $a \leq \bar{x} \leq b$.

Here, by virtue of formula (7), we have

$$\begin{aligned} P(a < x < b) &= P(a < x < \bar{x}) + P(\bar{x} < x < b) \\ &= P(a - \bar{x} < x - \bar{x} < 0) + P(0 < x - \bar{x} < b - \bar{x}) \\ &= \frac{1}{2}\left\{\Phi\left(\frac{\bar{x} - a}{Q_x}\right) + \Phi\left(\frac{b - \bar{x}}{Q_x}\right)\right\}. \end{aligned}$$

Third case: $a \leq b \leq \bar{x}$.

Here, we have

$$\begin{aligned} P(a < x < b) &= P(a < x < \bar{x}) - P(b < x < \bar{x}) \\ &= P(a - \bar{x} < x - \bar{x} < 0) - P(b - \bar{x} < x - \bar{x} < 0) \\ &= \frac{1}{2}\left\{\Phi\left(\frac{\bar{x} - a}{Q_x}\right) - \Phi\left(\frac{\bar{x} - b}{Q_x}\right)\right\}. \end{aligned}$$

The problem is thus solved for all three cases. We see that for a random variable distributed according to an arbitrary normal law, the table enables us to find the probability that this quantity fall into an arbitrary subinterval and by the same token characterizes completely its distribution law.

In order to see how the calculations are carried out in practice, we shall consider the following example.

EXAMPLE.   Firing is executed from the point O along the straight line OX.   The average range of the shell equals 1200 m.   Assuming that the range H is distributed according to a normal law with standard deviation 40 m., find what percentage of the shells overshoot the average range by 60 to 80 m.

In order that a shell have such a range, we must have $1260 < H < 1280$; applying the final formula in Problem III, first case, above, we find that

$$P(1260 < H < 1280) = \frac{1}{2}\left\{\varPhi\left(\frac{1280-1200}{40}\right) - \varPhi\left(\frac{1260-1200}{40}\right)\right\}$$

$$= \frac{1}{2}\{\varPhi(2) - \varPhi(1.5)\}.$$

From the table, we find that

$$\varPhi(2) \approx 0.955, \qquad \varPhi(1.5) \approx 0.866,$$

from which it follows that

$$P(1260 < H < 1280) \approx 0.044.$$

We see that a trifle more than 4% of the shells fired will have the indicated range.

CONCLUSION

During recent decades, the theory of probability has been transformed into one of the most rapidly developing mathematical sciences. New theoretical results reveal new possibilities for the utilization of the methods of probability theory in the natural sciences. A more detailed study of natural phenomena puts pressure at the same time on the theory of probability to search for new methods and new laws which are generated by chance. The theory of probability is one of those mathematical sciences which are not separated from life and from the problems of other sciences, but rather go hand in hand with the general development of the natural sciences and technology. The reader must not misunderstand what we have just asserted and think that the theory of probability is now transformed into only a support, an auxiliary tool, for the solution of practical problems. Not at all— during the last three decades the theory of probability has been transformed into a harmonious mathematical science with its own problems and methods of investigation. At the same time, it has turned out that the most important and natural problems of the theory of probability as a mathematical science assist in the solution of actual problems in the natural sciences.

The origin of the theory of probability goes back to the middle of the seventeenth century and is connected with the names Fermat (1601–1665), B. Pascal (1623–1662) and Huygens (1629–1695). In the works of these scholars, there appeared in embryo form the concepts of the probability of a stochastic event and of mathematical expectation (i.e., the expected or mean value) of a random variable. The point of departure for their investigations was problems connected with games of chance. However, the importance of new concepts for the study of nature was clear to them and Huygens, for example, in the collection *On Calculations in Games of Chance* wrote: "The reader will note that we are dealing not only with games, but also that the foundations of a very interesting and profound theory are being laid here." Among later scholars who exerted significant influence on the development of the theory of probability one must point out Jacob Bernoulli (1654–1705), whose name we have already met in the text of our book, De Moivre (1667–1754), Bayes (1702–1761), P. Laplace (1749–1827), Gauss (1777–1855), and Poisson (1781–1840).

The forceful development of probability theory is closely connected with the traditions and attainments of Russian science. At the time in the last century when probability theory went into eclipse in the West, in Russia the brilliant mathematician P. L. Chebyshev found a new means for its development—the over-all investigation of a sequence of independent random variables. Chebyshev himself, and his students A. A. Markov and A. M. Lyapunov, obtained fundamental results by this means (e.g., the law of large numbers and Lyapunov's theorem).

The reader is already acquainted with the law of large numbers, and our next problem now consists in giving an idea of the second important proposition of probability theory, which has been named Lyapunov's theorem—also called the central limit theorem of the theory of probability.

The reason for the great importance of this theorem is that a significant number of phenomena whose origin depends on chance proceed, in their fundamental behavior, according to the following scheme: the phenomenon under study is subjected to the action of an enormous number of independently acting stochastic causes each of which exerts only a negligibly small influence on the course of the phenomenon as a whole. The action of each of these causes is expressed by random variables $\xi_1, \xi_2, \ldots, \xi_n$, and their combined influence on the phenomenon equals the sum $s_n = \xi_1 + \xi_2 + \ldots + \xi_n$. Since the consideration of the influence of each of these causes (in other words, indication of the distribution function of the quantities ξ) and even the straightforward enumeration of them is practically impossible, it is clear to what extent it is important to evolve methods enabling one to study their combined action independently of the nature of each individual term. The usual methods of investigation are incapable of solving the problem posed—other methods must come to replace them, methods for which a large number of causes acting on the phenomenon would not be a hindrance, but would rather ease the solution of the posed problem. These methods should compensate for the insufficient knowledge of each of the individual acting causes by their large number—by their preponderance. The central limit theorem, established by the works, principally, of Academicians P. L. Chebyshev (1821–1894), A. A. Markov (1856–1922) and A. M. Lyapunov (1857–1918), asserts that if the acting causes $\xi_1, \xi_2, \ldots, \xi_n$ are mutually independent, if their number n is very large, and if the action of each of these causes in comparison with their total action is not large, then the law of distribution of the sum s_n can differ only insignificantly from a normal distribution law.

We introduce below examples of phenomena which proceed according to the scheme just described.

In firing from a cannon to a target, deviations of the point of impact of the shell from the point aimed at are unavoidable. This is the well known phenomenon of the dispersion of shells. Since the dispersion is the result of the action of an enormous number of independently acting factors (for example, the irregularities in the milling of the shell casing, the head of the shell, the variations in the density of the material of which the head of the shell is made, the minute variation in the quantity of explosive material in the various shells, the small errors, which are unnoticeable to the eye, in the aiming of the cannon, the minute variation in the composition of the atmosphere for the various firings, and many others), each of which influences only to an almost negligible degree the trajectory of the shell, then it follows from Lyapunov's theorem that it ought to be subject to a normal law. This situation is taken into consideration in the theory of artillery fire and is deemed fundamental in the evolution of firing rules.

When we carry out any observation with the purpose of measuring some physical constant, then an enormous number of factors each of which cannot be evaluated individually but which generates errors in measurements, unavoidably influence the result of our observation. In this number are included the errors in the condition of the measuring instrument, whose indications can vary without our knowing it under the influence of various atmospheric, temperature, mechanical and other causes. In this number are errors due to the observer, due to the peculiarities of his sight or hearing and due also to unknown effects depending on the psychological and physical condition of the observer. An actual error in measurement is thus the result of an enormous quantity of, so to speak, elementary errors which are negligible in magnitude, are mutually independent, and which depend on the case at hand. By virtue of Lyapunov's theorem, we can again expect that the errors in observation will be subject to a normal distribution law.

We can introduce as many such examples as we please: the position and velocity of the molecules of a gas, defined by a large number of collisions with other molecules; the quantity of diffused material; the deviation of the parts of a mechanism from a prescribed dimension in the mass production of mechanisms; the distribution of the growth of animals, plants or any of their organs, and so forth.

The perfection of physical statistics and also of a number of branches of technology placed before probability theory a large number of entirely new problems which do not fit into the framework of classical schemes. At the time when physics and technology were interested in *process* (i.e., in phenomena proceeding according to time), probability theory neither had general methods nor had it evolved particular schemes for the solution of problems which arose in the investigation of such phenomena. There arose a continual demand for the evolution of a general theory of *stochastic processes*, i.e., for a theory which would study random variables depending on one or several continuously varying parameters.

We shall enumerate several problems leading to the consideration of random variables whose variation proceeds with time. Let us imagine we have set for ourselves the goal of investigating the movement of some molecule of gas or liquid. This molecule collides at random moments with other molecules, in which connection it changes its speed and direction of motion. This variation in the condition of the molecule is subjected to the action of chance at every moment. Knowledge of a whole series of physical phenomena is required for its study, leading directly to a method of calculating the probability of the number of molecules that succeed in moving a certain distance in some interval of time. Thus, for instance, if two gases or two liquids are brought into contact, there then begins a mutual penetration of the molecules of one of the gases or liquids into the other; i.e., diffusion occurs. How rapidly does this diffusion process proceed, according to what laws, and when will the mixture of gases or liquids which is being formed become practically homogeneous? Answers to all these questions are given by the statistical theory of diffusion, on the basis of which lie the probabilistic considerations in the study of stochastic processes.

Obviously, an analogous problem arises also in chemistry when the process of the chemical interaction of substances is studied—i.e., the process of a chemical reaction. What portion of molecules has already entered into the reaction, how does the reaction proceed with time, when is the reaction practically complete?

A very important number of phenomena are due to radioactive decay. This phenomenon consists in that the atoms of a radioactive substance decompose, transforming into the atoms of another element. Each decomposition of an atom occurs in a moment, similar to an explosion with the emission of a certain quantity of energy. Numerous observations show that decompositions of atoms occur at random moments and independently of one another (with the condition that the mass of the radioactive substance is not too large). It is very essential in the investigation of the process of radioactive decay to determine what the probability is that after a prescribed interval of time a certain number of atoms will decompose. This is a typical problem in the theory of stochastic processes. Formally, if we are satisfied with only an explanation of the mathematical picture of the phenomenon, other phenomena proceed in exactly the same way: loads at a telephone station (i.e., the number of calls made at the telephone station by subscribers), Brownian movement, the breaking of thread on a weaving machine, and many others.

The origin of the general theory of stochastic processes was laid by the fundamental works of the Soviet mathematicians A. N. Kolmogorov and A. Ya. Khinchin at the beginning of the 1930's. Somewhat earlier, in the first decades of the present century, A. A. Markov began the study of sequences of dependent random variables, which sequences received the name Markov chains. The theory developed by him—at first only as a mathematical discipline—was transformed in the nineteen twenties, in the hands of physicists, into an active tool for the study of natural processes. From that time on, many scientists (S. N. Bernshtein, V. I. Romanovsky, A. N. Kolmogorov, J. Hadamard, M. Fréchet, W. Doeblin, J. Doob, W. Feller, and others) have made significant contributions to the theory of Markov chains.

In the 1920's, A. N. Kolmogorov, Ye. Ye. Slutsky, A. Ya. Khinchin, and Paul Lévy found a close connection between the theory of probability and the mathematical disciplines which are devoted to the study of sets and the general concept of function (the theory of sets and the theory of functions of a real variable). E. Borel had arrived at these same ideas somewhat earlier. The discovery of this connection proved to be extraordinarily fruitful and it was namely in this way that scientists succeeded in finding finally the solution of the classical problems posed by Chebyshev.

Finally, we must mention the works of S. N. Bernshtein, A. N. Kolmogorov, and von Mises on the construction of the logical harmonious theory of probability which was capable of dealing with various problems that had arisen previously in the natural sciences, in technology, and in other fields of knowledge.

In the contemporary turbulent development of probability theory an especially large role is played by science in the USSR, United States, France, Great Britain, Sweden, Japan, and Hungary. Moreover, interest in this scientific discipline has greatly grown in all countries largely under the influence of the persistent requirements of practice in its most variegated manifestations.

APPENDIX

Table of values of the function $\Phi(a)$

a	$\Phi(a)$	a	$\Phi(a)$	a	$\Phi(a)$	a	$\Phi(a)$	a	$\Phi(a)$
0.00	0.000	**0.60**	0.451	**1.20**	0.770	**1.80**	0.928	**2.40**	0.984
0.01	0.008	0.61	0.458	1.21	0.774	1.81	0.930	2.41	0.984
0.02	0.016	0.62	0.465	1.22	0.778	1.82	0.931	2.42	0.984
0.03	0.024	0.63	0.471	1.23	0.781	1.83	0.933	2.43	0.985
0.04	0.032	0.64	0.478	1.24	0.785	1.84	0.934	2.44	0.985
0.05	0.040	0.65	0.484	1.25	0.789	1.85	0.936	2.45	0.986
0.06	0.048	0.66	0.491	1.26	0.792	1.86	0.937	2.46	0.986
0.07	0.056	0.67	0.497	1.27	0.796	1.87	0.939	2.47	0.986
0.08	0.064	0.68	0.504	1.28	0.800	1.88	0.940	2.48	0.987
0.09	0.072	0.69	0.510	1.29	0.803	1.89	0.941	2.49	0.987
0.10	0.080	**0.70**	0.516	**1.30**	0.806	**1.90**	0.943	**2.50**	0.988
0.11	0.088	0.71	0.522	1.31	0.810	1.91	0.944	2.51	0.988
0.12	0.096	0.72	0.528	1.32	0.813	1.92	0.945	2.52	0.988
0.13	0.103	0.73	0.535	1.33	0.816	1.93	0.946	2.53	0.989
0.14	0.111	0.74	0.541	1.34	0.820	1.94	0.948	2.54	0.989
0.15	0.119	0.75	0.547	1.35	0.823	1.95	0.949	2.55	0.989
0.16	0.127	0.76	0.553	1.36	0.826	1.96	0.950	2.56	0.990
0.17	0.135	0.77	0.559	1.37	0.829	1.97	0.951	2.57	0.990
0.18	0.143	0.78	0.565	1.38	0.832	1.98	0.952	2.58	0.990
0.19	0.151	0.79	0.570	1.39	0.835	1.99	0.953	2.59	0.990
0.20	0.159	**0.80**	0.576	**1.40**	0.838	**2.00**	0.955	**2.60**	0.991
0.21	0.166	0.81	0.582	1.41	0.841	2.01	0.956	2.61	0.991
0.22	0.174	0.82	0.588	1.42	0.844	2.02	0.957	2.62	0.991
0.23	0.182	0.83	0.593	1.43	0.847	2.03	0.958	2.63	0.991
0.24	0.190	0.84	0.599	1.44	0.850	2.04	0.959	2.64	0.992
0.25	0.197	0.85	0.605	1.45	0.853	2.05	0.960	2.65	0.992
0.26	0.205	0.86	0.610	1.46	0.856	2.06	0.961	2.66	0.992
0.27	0.213	0.87	0.616	1.47	0.858	2.07	0.962	2.67	0.992
0.28	0.221	0.88	0.621	1.48	0.861	2.08	0.962	2.68	0.993
0.29	0.228	0.89	0.627	1.49	0.864	2.09	0.963	2.69	0.993
0.30	0.236	**0.90**	0.632	**1.50**	0.866	**2.10**	0.964	**2.70**	0.993
0.31	0.243	0.91	0.637	1.51	0.867	2.11	0.965	2.72	0.993
0.32	0.251	0.92	0.642	1.52	0.871	2.12	0.966	2.74	0.994
0.33	0.259	0.93	0.648	1.53	0.874	2.13	0.967	2.76	0.994
0.34	0.266	0.94	0.653	1.54	0.876	2.14	0.968	2.78	0.995
0.35	0.274	0.95	0.658	1.55	0.879	2.15	0.968		
0.36	0.281	0.96	0.663	1.56	0.881	2.16	0.969	**2.80**	0.995
0.37	0.289	0.97	0.668	1.57	0.884	2.17	0.970	2.82	0.995
0.38	0.296	0.98	0.673	1.58	0.886	2.18	0.971	2.84	0.995
0.39	0.303	0.99	0.678	1.59	0.888	2.19	0.971	2.86	0.996
								2.88	0.996
0.40	0.311	**1.00**	0.683	**1.60**	0.890	**2.20**	0.972		
0.41	0.318	1.01	0.688	1.61	0.893	2.21	0.973	**2.90**	0.996
0.42	0.326	1.02	0.692	1.62	0.895	2.22	0.974	2.92	0.996
0.43	0.333	1.03	0.697	1.63	0.897	2.23	0.974	2.94	0.997
0.44	0.340	1.04	0.702	1.64	0.899	2.24	0.975	2.96	0.997
0.45	0.347	1.05	0.706	1.65	0.901	2.25	0.976	2.98	0.997
0.46	0.354	1.06	0.711	1.66	0.903	2.26	0.976		
0.47	0.362	1.07	0.715	1.67	0.905	2.27	0.977	**3.00**	0.997
0.48	0.369	1.08	0.720	1.68	0.907	2.28	0.977	3.10	0.998
0.49	0.376	1.09	0.724	1.69	0.099	2.29	0.978	3.20	0.999
								3.30	0.999
								3.40	0.999
0.50	0.383	**1.10**	0.729	**1.70**	0.911	**2.30**	0.979		
0.51	0.390	1.11	0.733	1.71	0.913	2.31	0.979	**3.50**	0.9995
0.52	0.397	1.12	0.737	1.72	0.915	2.32	0.980	3.60	0.9997
0.53	0.404	1.13	0.742	1.73	0.916	2.33	0.980	3.70	0.9998
0.54	0.411	1.14	0.746	1.74	0.918	2.34	0.981	3.80	0.99986
0.55	0.418	1.15	0.750	1.75	0.920	2.35	0.981	3.90	0.99990
0.56	0.425	1.16	0.754	1.76	0.922	2.36	0.982		
0.57	0.431	1.17	0.758	1.77	0.923	2.37	0.982	**4.00**	0.99994
0.58	0.438	1.18	0.762	1.78	0.925	2.38	0.983	**5.00**	0.99999994
0.59	0.445	1.19	0.766	1.79	0.927	2.39	0.983		

BIBLIOGRAPHY

ARLEY, N. and BUCH, K., *Introduction to the Theory of Probability and Statistics*. John Wiley & Sons, New York, 1950.

——, *Introduction to the Theory of Probability and Mathematical Statistics*, IIL, Moscow, 1951 (Russian).

BERNSHTEIN, S. N., *Theory of Probability*, 4th ed. Gostehizdat, Moscow–Leningrad, 1946 (Russian).

BIZLEY, M., *Probability: An Intermediate Textbook*. Cambridge University Press, Cambridge, 1957.

BURINGTON, R. and MAY, D., *Handbook of Probability and Statistics with Tables*. Handbook Publishers, Sandusky, 1953.

BURNSIDE, W., *Theory of Probability*. Dover Publications, New York, 1959.

CASTELNUOVO, G., *Calcolo delle probabilità*, 2nd ed. Bologna, 1926–1928.

CRAMÉR, H., *The Elements of Probability Theory and some of its Applications*. John Wiley & Sons, New York, 1955.

——, *Sannolikhetskalkylen och några av dess användningar*, Uppsala.

——, *Random Variables and Probability Distributions*, Cambridge University Press, Cambridge, 1937.

——, *Random Variables and Probability Distributions*, IIL, Moscow, 1947 (Russian).

DOOB, J., *Stochastic Processes*. John Wiley & Sons, New York, 1953.

FELLER, W., *An Introduction to Probability Theory and Its Applications*. John Wiley & Sons, New York, 1950, 1957.

——, *An Introduction to Probability Theory and its Applications*, IIL, Moscow, 1952 (Russian).

FISZ, M., *Rachunek prawdopodobieństwa i statystyka matematyczna*. Państwowe Wydawnictow Naukowe, Warsaw, 1954.

——, *Wahrscheinlichkeitsrechnung und mathematische Statistik*. Deutscher Verlag d. Wissenschaften, Berlin, 1958.

FORTET, R., *Calcul des probabilités*, Centre National de la Recherche Scientifique, Paris, 1950.

FRY, T. C., *Probability and its Engineering Uses*. D. Van Nostrand, New York, 1928.

GNEDENKO, B. V., *Course in the Theory of Probability*, 2nd ed. GITTL, Moscow, 1954 (Russian).

GNEDENKO, B., *Theory of Probability*. Chelsea, New York (in' preparation).

GOLDBERG, S., *Probability: An Introduction*. Prentice-Hall, Englewood Cliffs, 1960.

HOSTINSKY, B., *Méthodes générales du calcul des probabilités*. Mém. Sci. Math., Vol. 52, Paris, 1931.

ITÔ, K., *Theory of Probability*. Iwanami Shôten, Tokyo, 1953 (Japanese).

KAWADA, Y., *The Theory of Probability*. Kyôritsusha, Tokyo, 1952 (Japanese).

KEMENY, J. G., MIRKIL, H., SNELL, J. L. and THOMPSON, G. L., *Finite Mathematical Structures*. Prentice-Hall, Englewood Cliffs, 1959.

KEMENY, J. G., SNELL, J. L. and THOMPSON, G. L., *Introduction to Finite Mathematics*. Prentice-Hall, Englewood Cliffs, 1957.

KHINCHIN, A. YA., *Asymptotische Gesetze der Wahrscheinlichkeitsrechnung*. Chelsea, New York, 1948 and Springer, Berlin, 1933.

——, *Asymptotic Laws in the Theory of Probability*. GITTL, Moscow–Leningrad, 1936 (Russian).

KNEALE, W., *Probability and Induction*. Clarendon Press, Oxford, 1949.

KOLMOGOROV, A., *Foundations of the Theory of Probability*. Chelsea, New York, 1956.

——, *Grundbegriffe der Wahrscheinlichkeitsrechnung*. Springer, Berlin, 1933.

——, *Foundations of the Theory of Probability*. GITTL, Moscow–Leningrad, 1936 (Russian).

LÉVY, P., *Théorie de l'addition des variables aléatoires*. Gauthier-Villars, Paris, 1937.

——, *Calcul des probabilités*. Gauthiers-Villars, Paris, 1954.

LINDGREN, B. W. and McELRATH, G. W., *Introduction to Probability and Statistics*. Macmillan, New York, 1959.

LOÈVE, M., *Probability Theory*, 2nd ed. D. Van Nostrand, Princeton, 1960.

——, *Probability Theory: Foundations, Random Sequences*. D. Van Nostrand, New York, 1955.

MACK, S. F., *Elementary Statistics*. Holt, Rinehart and Winston, New York, 1960.

MISES, R. V., *Wahrscheinlichkeitsrechnung*. Deuticke, Leipzig–Wien, 1931.

MUNROE, M., *The Theory of Probability*. McGraw-Hill, New York, 1951.

NEYMAN, J., *First Course in Probability and Statistics.* Holt, Rinehart and Winston, New York, 1950.

PARZEN, E., *Modern Probability Theory and its Applications.* John Wiley & Sons, New York, 1960.

RÉNYI, A., *The Calculus of Probabilities.* Tanköyvkiadó, Budapest, 1954 (Hungarian).

RIORDAN, J., *An Introduction to Combinatorial Analysis.* John Wiley & Sons, New York, 1958.

RUMSHISKY, L., *Elements of the Theory of Probability.* Fizmatgiz, Moscow, 1960 (Russian).

SARYMSAKOV, T. A., *Elements of the Theory of Markov Processes.* GITTL, Moscow, 1954 (Russian).

SCHLAIFER, R., *Probability and Statistics for Business Decisions; an Introduction to Managerial Economics under Uncertainty.* McGraw-Hill, New York, 1959.

TODHUNTER, I., *A History of the Mathematical Theory of Probability.* Chelsea, New York, 1949.

TORNIER, E., *Wahrscheinlichkeitsrechnung und allgemeine Integrationstheorie.* Teubner, Leipzig–Berlin, 1936.

USPENSKY, J., *Introduction to Mathematical Probability.* McGraw-Hill, New York, 1937.

VENTSEL, YE., *Theory of Probability.* Gos. Izdat. Fiz.-Mat. Lit., Moscow, 1958 (Russian).

——, *Theory of Probability.* Chelsea, New York, 1961.

——, *Theory of Probability,* Pergamon, London (in preparation).

WOODWARD, P., *Probability and Information Theory, with Applications to Radar.* Pergamon Press, London, 1953.

YAGLOM, A. and YAGLOM, I., *Probability and Information,* 2nd ed. Fizmatgiz, Moscow, 1960 (Russian).

——, *Probability and Information.* Pergamon, London (in preparation).

INDEX

A CATALOG OF SELECTED
DOVER BOOKS
IN SCIENCE AND MATHEMATICS

Astronomy

BURNHAM'S CELESTIAL HANDBOOK, Robert Burnham, Jr. Thorough guide to the stars beyond our solar system. Exhaustive treatment. Alphabetical by constellation: Andromeda to Cetus in Vol. 1; Chamaeleon to Orion in Vol. 2; and Pavo to Vulpecula in Vol. 3. Hundreds of illustrations. Index in Vol. 3. 2,000pp. 6⅛ x 9¼.

Vol. I: 0-486-23567-X
Vol. II: 0-486-23568-8
Vol. III: 0-486-23673-0

EXPLORING THE MOON THROUGH BINOCULARS AND SMALL TELE-SCOPES, Ernest H. Cherrington, Jr. Informative, profusely illustrated guide to locating and identifying craters, rills, seas, mountains, other lunar features. Newly revised and updated with special section of new photos. Over 100 photos and diagrams. 240pp. 8¼ x 11. 0-486-24491-1

THE EXTRATERRESTRIAL LIFE DEBATE, 1750–1900, Michael J. Crowe. First detailed, scholarly study in English of the many ideas that developed from 1750 to 1900 regarding the existence of intelligent extraterrestrial life. Examines ideas of Kant, Herschel, Voltaire, Percival Lowell, many other scientists and thinkers. 16 illustrations. 704pp. 5⅜ x 8½. 0-486-40675-X

THEORIES OF THE WORLD FROM ANTIQUITY TO THE COPERNICAN REVOLUTION, Michael J. Crowe. Newly revised edition of an accessible, enlightening book re-creates the change from an earth-centered to a sun-centered conception of the solar system. 242pp. 5⅜ x 8½. 0-486-41444-2

ARISTARCHUS OF SAMOS: The Ancient Copernicus, Sir Thomas Heath. Heath's history of astronomy ranges from Homer and Hesiod to Aristarchus and includes quotes from numerous thinkers, compilers, and scholasticists from Thales and Anaximander through Pythagoras, Plato, Aristotle, and Heraclides. 34 figures. 448pp. 5⅜ x 8½.
0-486-43886-4

A COMPLETE MANUAL OF AMATEUR ASTRONOMY: TOOLS AND TECHNIQUES FOR ASTRONOMICAL OBSERVATIONS, P. Clay Sherrod with Thomas L. Koed. Concise, highly readable book discusses: selecting, setting up and maintaining a telescope; amateur studies of the sun; lunar topography and occultations; observations of Mars, Jupiter, Saturn, the minor planets and the stars; an introduction to photoelectric photometry; more. 1981 ed. 124 figures. 25 halftones. 37 tables. 335pp. 6½ x 9¼. 0-486-42820-8

AMATEUR ASTRONOMER'S HANDBOOK, J. B. Sidgwick. Timeless, comprehensive coverage of telescopes, mirrors, lenses, mountings, telescope drives, micrometers, spectroscopes, more. 189 illustrations. 576pp. 5⅝ x 8¼. (Available in U.S. only.)
0-486-24034-7

STAR LORE: Myths, Legends, and Facts, William Tyler Olcott. Captivating retellings of the origins and histories of ancient star groups include Pegasus, Ursa Major, Pleiades, signs of the zodiac, and other constellations. "Classic."—Sky & Telescope. 58 illustrations. 544pp. 5⅜ x 8½. 0-486-43581-4

Chemistry

THE SCEPTICAL CHYMIST: THE CLASSIC 1661 TEXT, Robert Boyle. Boyle defines the term "element," asserting that all natural phenomena can be explained by the motion and organization of primary particles. 1911 ed. viii+232pp. $5^3/_8$ x $8^1/_2$.
0-486-42825-7

RADIOACTIVE SUBSTANCES, Marie Curie. Here is the celebrated scientist's doctoral thesis, the prelude to her receipt of the 1903 Nobel Prize. Curie discusses establishing atomic character of radioactivity found in compounds of uranium and thorium; extraction from pitchblende of polonium and radium; isolation of pure radium chloride; determination of atomic weight of radium; plus electric, photographic, luminous, heat, color effects of radioactivity. ii+94pp. $5^3/_8$ x $8^1/_2$.
0-486-42550-9

CHEMICAL MAGIC, Leonard A. Ford. Second Edition, Revised by E. Winston Grundmeier. Over 100 unusual stunts demonstrating cold fire, dust explosions, much more. Text explains scientific principles and stresses safety precautions. 128pp. $5^3/_8$ x $8^1/_2$.
0-486-67628-5

MOLECULAR THEORY OF CAPILLARITY, J. S. Rowlinson and B. Widom. History of surface phenomena offers critical and detailed examination and assessment of modern theories, focusing on statistical mechanics and application of results in mean-field approximation to model systems. 1989 edition. 352pp. $5^3/_8$ x $8^1/_2$.
0-486-42544-4

CHEMICAL AND CATALYTIC REACTION ENGINEERING, James J. Carberry. Designed to offer background for managing chemical reactions, this text examines behavior of chemical reactions and reactors; fluid-fluid and fluid-solid reaction systems; heterogeneous catalysis and catalytic kinetics; more. 1976 edition. 672pp. $6^1/_8$ x $9^1/_4$.
0-486-41736-0 $31.95

ELEMENTS OF CHEMISTRY, Antoine Lavoisier. Monumental classic by founder of modern chemistry in remarkable reprint of rare 1790 Kerr translation. A must for every student of chemistry or the history of science. 539pp. $5^3/_8$ x $8^1/_2$.
0-486-64624-6

MOLECULES AND RADIATION: An Introduction to Modern Molecular Spectroscopy. Second Edition, Jeffrey I. Steinfeld. This unified treatment introduces upper-level undergraduates and graduate students to the concepts and the methods of molecular spectroscopy and applications to quantum electronics, lasers, and related optical phenomena. 1985 edition. 512pp. $5^3/_8$ x $8^1/_2$.
0-486-44152-0

A SHORT HISTORY OF CHEMISTRY, J. R. Partington. Classic exposition explores origins of chemistry, alchemy, early medical chemistry, nature of atmosphere, theory of valency, laws and structure of atomic theory, much more. 428pp. $5^3/_8$ x $8^1/_2$. (Available in U.S. only.)
0-486-65977-1

GENERAL CHEMISTRY, Linus Pauling. Revised 3rd edition of classic first-year text by Nobel laureate. Atomic and molecular structure, quantum mechanics, statistical mechanics, thermodynamics correlated with descriptive chemistry. Problems. 992pp. $5^3/_8$ x $8^1/_2$.
0-486-65622-5

ELECTRON CORRELATION IN MOLECULES, S. Wilson. This text addresses one of theoretical chemistry's central problems. Topics include molecular electronic structure, independent electron models, electron correlation, the linked diagram theorem, and related topics. 1984 edition. 304pp. $5^3/_8$ x $8^1/_2$.
0-486-45879-2

Engineering

DE RE METALLICA, Georgius Agricola. The famous Hoover translation of greatest treatise on technological chemistry, engineering, geology, mining of early modern times (1556). All 289 original woodcuts. 638pp. 6³/₄ x 11. 0-486-60006-8

FUNDAMENTALS OF ASTRODYNAMICS, Roger Bate et al. Modern approach developed by U.S. Air Force Academy. Designed as a first course. Problems, exercises. Numerous illustrations. 455pp. 5³/₈ x 8¹/₂. 0-486-60061-0

DYNAMICS OF FLUIDS IN POROUS MEDIA, Jacob Bear. For advanced students of ground water hydrology, soil mechanics and physics, drainage and irrigation engineering and more. 335 illustrations. Exercises, with answers. 784pp. 6¹/₈ x 9¹/₄. 0-486-65675-6

THEORY OF VISCOELASTICITY (SECOND EDITION), Richard M. Christensen. Complete consistent description of the linear theory of the viscoelastic behavior of materials. Problem-solving techniques discussed. 1982 edition. 29 figures. xiv+364pp. 6¹/₈ x 9¹/₄. 0-486-42880-X

MECHANICS, J. P. Den Hartog. A classic introductory text or refresher. Hundreds of applications and design problems illuminate fundamentals of trusses, loaded beams and cables, etc. 334 answered problems. 462pp. 5³/₈ x 8¹/₂. 0-486-60754-2

MECHANICAL VIBRATIONS, J. P. Den Hartog. Classic textbook offers lucid explanations and illustrative models, applying theories of vibrations to a variety of practical industrial engineering problems. Numerous figures. 233 problems, solutions. Appendix. Index. Preface. 436pp. 5³/₈ x 8¹/₂. 0-486-64785-4

STRENGTH OF MATERIALS, J. P. Den Hartog. Full, clear treatment of basic material (tension, torsion, bending, etc.) plus advanced material on engineering methods, applications. 350 answered problems. 323pp. 5³/₈ x 8¹/₂. 0-486-60755-0

A HISTORY OF MECHANICS, René Dugas. Monumental study of mechanical principles from antiquity to quantum mechanics. Contributions of ancient Greeks, Galileo, Leonardo, Kepler, Lagrange, many others. 671pp. 5³/₈ x 8¹/₂. 0-486-65632-2

STABILITY THEORY AND ITS APPLICATIONS TO STRUCTURAL MECHANICS, Clive L. Dym. Self-contained text focuses on Koiter postbuckling analyses, with mathematical notions of stability of motion. Basing minimum energy principles for static stability upon dynamic concepts of stability of motion, it develops asymptotic buckling and postbuckling analyses from potential energy considerations, with applications to columns, plates, and arches. 1974 ed. 208pp. 5³/₈ x 8¹/₂. 0-486-42541-X

BASIC ELECTRICITY, U.S. Bureau of Naval Personnel. Originally a training course; best nontechnical coverage. Topics include batteries, circuits, conductors, AC and DC, inductance and capacitance, generators, motors, transformers, amplifiers, etc. Many questions with answers. 349 illustrations. 1969 edition. 448pp. 6¹/₂ x 9¹/₄. 0-486-20973-3

ROCKETS, Robert Goddard. Two of the most significant publications in the history of rocketry and jet propulsion: "A Method of Reaching Extreme Altitudes" (1919) and "Liquid Propellant Rocket Development" (1936). 128pp. $5\frac{3}{8}$ x $8\frac{1}{2}$.　　0-486-42537-1

STATISTICAL MECHANICS: PRINCIPLES AND APPLICATIONS, Terrell L. Hill. Standard text covers fundamentals of statistical mechanics, applications to fluctuation theory, imperfect gases, distribution functions, more. 448pp. $5\frac{3}{8}$ x $8\frac{1}{2}$.　　0-486-65390-0

ENGINEERING AND TECHNOLOGY 1650–1750: ILLUSTRATIONS AND TEXTS FROM ORIGINAL SOURCES, Martin Jensen. Highly readable text with more than 200 contemporary drawings and detailed engravings of engineering projects dealing with surveying, leveling, materials, hand tools, lifting equipment, transport and erection, piling, bailing, water supply, hydraulic engineering, and more. Among the specific projects outlined-transporting a 50-ton stone to the Louvre, erecting an obelisk, building timber locks, and dredging canals. 207pp. $8\frac{3}{8}$ x $11\frac{1}{4}$.　　0-486-42232-1

THE VARIATIONAL PRINCIPLES OF MECHANICS, Cornelius Lanczos. Graduate level coverage of calculus of variations, equations of motion, relativistic mechanics. First inexpensive paperbound edition of classic treatise. Index. Bibliography. 418pp. $5\frac{3}{8}$ x $8\frac{1}{2}$.　　0-486-65067-7

PROTECTION OF ELECTRONIC CIRCUITS FROM OVERVOLTAGES, Ronald B. Standler. Five-part treatment presents practical rules and strategies for circuits designed to protect electronic systems from damage by transient overvoltages. 1989 ed. xxiv+434pp. $6\frac{1}{8}$ x $9\frac{1}{4}$.　　0-486-42552-5

ROTARY WING AERODYNAMICS, W. Z. Stepniewski. Clear, concise text covers aerodynamic phenomena of the rotor and offers guidelines for helicopter performance evaluation. Originally prepared for NASA. 537 figures. 640pp. $6\frac{1}{8}$ x $9\frac{1}{4}$. 0-486-64647-5

INTRODUCTION TO SPACE DYNAMICS, William Tyrrell Thomson. Comprehensive, classic introduction to space-flight engineering for advanced undergraduate and graduate students. Includes vector algebra, kinematics, transformation of coordinates. Bibliography. Index. 352pp. $5\frac{3}{8}$ x $8\frac{1}{2}$.　　0-486-65113-4

HISTORY OF STRENGTH OF MATERIALS, Stephen P. Timoshenko. Excellent historical survey of the strength of materials with many references to the theories of elasticity and structure. 245 figures. 452pp. $5\frac{3}{8}$ x $8\frac{1}{2}$.　　0-486-61187-6

ANALYTICAL FRACTURE MECHANICS, David J. Unger. Self-contained text supplements standard fracture mechanics texts by focusing on analytical methods for determining crack-tip stress and strain fields. 336pp. $6\frac{1}{8}$ x $9\frac{1}{4}$.　　0-486-41737-9

STATISTICAL MECHANICS OF ELASTICITY, J. H. Weiner. Advanced, self-contained treatment illustrates general principles and elastic behavior of solids. Part 1, based on classical mechanics, studies thermoelastic behavior of crystalline and polymeric solids. Part 2, based on quantum mechanics, focuses on interatomic force laws, behavior of solids, and thermally activated processes. For students of physics and chemistry and for polymer physicists. 1983 ed. 96 figures. 496pp. $5\frac{3}{8}$ x $8\frac{1}{2}$.　　0-486-42260-7

Mathematics

FUNCTIONAL ANALYSIS (Second Corrected Edition), George Bachman and Lawrence Narici. Excellent treatment of subject geared toward students with background in linear algebra, advanced calculus, physics and engineering. Text covers introduction to inner-product spaces, normed, metric spaces, and topological spaces; complete orthonormal sets, the Hahn-Banach Theorem and its consequences, and many other related subjects. 1966 ed. 544pp. 6⅛ x 9¼.　　　　　　　　　　　　　　　　0-486-40251-7

DIFFERENTIAL MANIFOLDS, Antoni A. Kosinski. Introductory text for advanced undergraduates and graduate students presents systematic study of the topological structure of smooth manifolds, starting with elements of theory and concluding with method of surgery. 1993 edition. 288pp. 5⅜ x 8½.　　　　　　　　　　　　0-486-46244-7

VECTOR AND TENSOR ANALYSIS WITH APPLICATIONS, A. I. Borisenko and I. E. Tarapov. Concise introduction. Worked-out problems, solutions, exercises. 257pp. 5⅜ x 8¼.　　　　　　　　　　　　　　　　　　　　　　0-486-63833-2

AN INTRODUCTION TO ORDINARY DIFFERENTIAL EQUATIONS, Earl A. Coddington. A thorough and systematic first course in elementary differential equations for undergraduates in mathematics and science, with many exercises and problems (with answers). Index. 304pp. 5⅜ x 8½.　　　　　　　　　　　　0-486-65942-9

FOURIER SERIES AND ORTHOGONAL FUNCTIONS, Harry F. Davis. An incisive text combining theory and practical example to introduce Fourier series, orthogonal functions and applications of the Fourier method to boundary-value problems. 570 exercises. Answers and notes. 416pp. 5⅜ x 8½.　　　　　　　　　0-486-65973-9

COMPUTABILITY AND UNSOLVABILITY, Martin Davis. Classic graduate-level introduction to theory of computability, usually referred to as theory of recurrent functions. New preface and appendix. 288pp. 5⅜ x 8½.　　　　　0-486-61471-9

AN INTRODUCTION TO MATHEMATICAL ANALYSIS, Robert A. Rankin. Dealing chiefly with functions of a single real variable, this text by a distinguished educator introduces limits, continuity, differentiability, integration, convergence of infinite series, double series, and infinite products. 1963 edition. 624pp. 5⅜ x 8½.　　0-486-46251-X

METHODS OF NUMERICAL INTEGRATION (SECOND EDITION), Philip J. Davis and Philip Rabinowitz. Requiring only a background in calculus, this text covers approximate integration over finite and infinite intervals, error analysis, approximate integration in two or more dimensions, and automatic integration. 1984 edition. 624pp. 5⅜ x 8½.　　　　　　　　　　　　　　　　　　　　　0-486-45339-1

INTRODUCTION TO LINEAR ALGEBRA AND DIFFERENTIAL EQUATIONS, John W. Dettman. Excellent text covers complex numbers, determinants, orthonormal bases, Laplace transforms, much more. Exercises with solutions. Undergraduate level. 416pp. 5⅜ x 8½.　　　　　　　　　　　　　　　　　0-486-65191-6

RIEMANN'S ZETA FUNCTION, H. M. Edwards. Superb, high-level study of landmark 1859 publication entitled "On the Number of Primes Less Than a Given Magnitude" traces developments in mathematical theory that it inspired. xiv+315pp. 5⅜ x 8½.

0-486-41740-9

CALCULUS OF VARIATIONS WITH APPLICATIONS, George M. Ewing. Applications-oriented introduction to variational theory develops insight and promotes understanding of specialized books, research papers. Suitable for advanced undergraduate/graduate students as primary, supplementary text. 352pp. 5³/₈ x 8¹/₂.
0-486-64856-7

MATHEMATICIAN'S DELIGHT, W. W. Sawyer. "Recommended with confidence" by *The Times Literary Supplement,* this lively survey was written by a renowned teacher. It starts with arithmetic and algebra, gradually proceeding to trigonometry and calculus. 1943 edition. 240pp. 5³/₈ x 8¹/₂.
0-486-46240-4

ADVANCED EUCLIDEAN GEOMETRY, Roger A. Johnson. This classic text explores the geometry of the triangle and the circle, concentrating on extensions of Euclidean theory, and examining in detail many relatively recent theorems. 1929 edition. 336pp. 5³/₈ x 8¹/₂.
0-486-46237-4

COUNTEREXAMPLES IN ANALYSIS, Bernard R. Gelbaum and John M. H. Olmsted. These counterexamples deal mostly with the part of analysis known as "real variables." The first half covers the real number system, and the second half encompasses higher dimensions. 1962 edition. xxiv+198pp. 5³/₈ x 8¹/₂.
0-486-42875-3

CATASTROPHE THEORY FOR SCIENTISTS AND ENGINEERS, Robert Gilmore. Advanced-level treatment describes mathematics of theory grounded in the work of Poincaré, R. Thom, other mathematicians. Also important applications to problems in mathematics, physics, chemistry and engineering. 1981 edition. References. 28 tables. 397 black-and-white illustrations. xvii + 666pp. 6¹/₈ x 9¹/₄.
0-486-67539-4

COMPLEX VARIABLES: Second Edition, Robert B. Ash and W. P. Novinger. Suitable for advanced undergraduates and graduate students, this newly revised treatment covers Cauchy theorem and its applications, analytic functions, and the prime number theorem. Numerous problems and solutions. 2004 edition. 224pp. 6¹/₂ x 9¹/₄.
0-486-46250-1

NUMERICAL METHODS FOR SCIENTISTS AND ENGINEERS, Richard Hamming. Classic text stresses frequency approach in coverage of algorithms, polynomial approximation, Fourier approximation, exponential approximation, other topics. Revised and enlarged 2nd edition. 721pp. 5³/₈ x 8¹/₂.
0-486-65241-6

INTRODUCTION TO NUMERICAL ANALYSIS (2nd Edition), F. B. Hildebrand. Classic, fundamental treatment covers computation, approximation, interpolation, numerical differentiation and integration, other topics. 150 new problems. 669pp. 5³/₈ x 8¹/₂.
0-486-65363-3

MARKOV PROCESSES AND POTENTIAL THEORY, Robert M. Blumental and Ronald K. Getoor. This graduate-level text explores the relationship between Markov processes and potential theory in terms of excessive functions, multiplicative functionals and subprocesses, additive functionals and their potentials, and dual processes. 1968 edition. 320pp. 5³/₈ x 8¹/₂.
0-486-46263-3

ABSTRACT SETS AND FINITE ORDINALS: An Introduction to the Study of Set Theory, G. B. Keene. This text unites logical and philosophical aspects of set theory in a manner intelligible to mathematicians without training in formal logic and to logicians without a mathematical background. 1961 edition. 112pp. 5³/₈ x 8¹/₂.
0-486-46249-8

INTRODUCTORY REAL ANALYSIS, A.N. Kolmogorov, S. V. Fomin. Translated by Richard A. Silverman. Self-contained, evenly paced introduction to real and functional analysis. Some 350 problems. 403pp. 5³/₈ x 8¹/₂.　　　　　　0-486-61226-0

APPLIED ANALYSIS, Cornelius Lanczos. Classic work on analysis and design of finite processes for approximating solution of analytical problems. Algebraic equations, matrices, harmonic analysis, quadrature methods, much more. 559pp. 5³/₈ x 8¹/₂.　0-486-65656-X

AN INTRODUCTION TO ALGEBRAIC STRUCTURES, Joseph Landin. Superb self-contained text covers "abstract algebra": sets and numbers, theory of groups, theory of rings, much more. Numerous well-chosen examples, exercises. 247pp. 5³/₈ x 8¹/₂.
0-486-65940-2

QUALITATIVE THEORY OF DIFFERENTIAL EQUATIONS, V. V. Nemytskii and V.V. Stepanov. Classic graduate-level text by two prominent Soviet mathematicians covers classical differential equations as well as topological dynamics and ergodic theory. Bibliographies. 523pp. 5³/₈ x 8¹/₂.　　　　　　　　　　　0-486-65954-2

THEORY OF MATRICES, Sam Perlis. Outstanding text covering rank, nonsingularity and inverses in connection with the development of canonical matrices under the relation of equivalence, and without the intervention of determinants. Includes exercises. 237pp. 5³/₈ x 8¹/₂.　　　　　　　　　　　　　　　　　　　　0-486-66810-X

INTRODUCTION TO ANALYSIS, Maxwell Rosenlicht. Unusually clear, accessible coverage of set theory, real number system, metric spaces, continuous functions, Riemann integration, multiple integrals, more. Wide range of problems. Undergraduate level. Bibliography. 254pp. 5³/₈ x 8¹/₂.　　　　　　　　　　　　　0-486-65038-3

MODERN NONLINEAR EQUATIONS, Thomas L. Saaty. Emphasizes practical solution of problems; covers seven types of equations. ". . . a welcome contribution to the existing literature. . . ."—*Math Reviews*. 490pp. 5³/₈ x 8¹/₂.　　　　0-486-64232-1

MATRICES AND LINEAR ALGEBRA, Hans Schneider and George Phillip Barker. Basic textbook covers theory of matrices and its applications to systems of linear equations and related topics such as determinants, eigenvalues and differential equations. Numerous exercises. 432pp. 5³/₈ x 8¹/₂.　　　　　　　　　　　　0-486-66014-1

LINEAR ALGEBRA, Georgi E. Shilov. Determinants, linear spaces, matrix algebras, similar topics. For advanced undergraduates, graduates. Silverman translation. 387pp. 5³/₈ x 8¹/₂.　　　　　　　　　　　　　　　　　　　　0-486-63518-X

MATHEMATICAL METHODS OF GAME AND ECONOMIC THEORY: Revised Edition, Jean-Pierre Aubin. This text begins with optimization theory and convex analysis, followed by topics in game theory and mathematical economics, and concluding with an introduction to nonlinear analysis and control theory. 1982 edition. 656pp. 6¹/₈ x 9¹/₄.
0-486-46265-X

SET THEORY AND LOGIC, Robert R. Stoll. Lucid introduction to unified theory of mathematical concepts. Set theory and logic seen as tools for conceptual understanding of real number system. 496pp. 5³/₈ x 8¹/₄.　　　　　　　　　0-486-63829-4

TENSOR CALCULUS, J.L. Synge and A. Schild. Widely used introductory text covers spaces and tensors, basic operations in Riemannian space, non-Riemannian spaces, etc. 324pp. $5^3/_8$ x $8^1/_4$. 0-486-63612-7

ORDINARY DIFFERENTIAL EQUATIONS, Morris Tenenbaum and Harry Pollard. Exhaustive survey of ordinary differential equations for undergraduates in mathematics, engineering, science. Thorough analysis of theorems. Diagrams. Bibliography. Index. 818pp. $5^3/_8$ x $8^1/_2$. 0-486-64940-7

INTEGRAL EQUATIONS, F. G. Tricomi. Authoritative, well-written treatment of extremely useful mathematical tool with wide applications. Volterra Equations, Fredholm Equations, much more. Advanced undergraduate to graduate level. Exercises. Bibliography. 238pp. $5^3/_8$ x $8^1/_2$. 0-486-64828-1

FOURIER SERIES, Georgi P. Tolstov. Translated by Richard A. Silverman. A valuable addition to the literature on the subject, moving clearly from subject to subject and theorem to theorem. 107 problems, answers. 336pp. $5^3/_8$ x $8^1/_2$. 0-486-63317-9

INTRODUCTION TO MATHEMATICAL THINKING, Friedrich Waismann. Examinations of arithmetic, geometry, and theory of integers; rational and natural numbers; complete induction; limit and point of accumulation; remarkable curves; complex and hypercomplex numbers, more. 1959 ed. 27 figures. xii+260pp. $5^3/_8$ x $8^1/_2$. 0-486-42804-8

THE RADON TRANSFORM AND SOME OF ITS APPLICATIONS, Stanley R. Deans. Of value to mathematicians, physicists, and engineers, this excellent introduction covers both theory and applications, including a rich array of examples and literature. Revised and updated by the author. 1993 edition. 304pp. $6^1/_8$ x $9^1/_4$. 0-486-46241-2

CALCULUS OF VARIATIONS, Robert Weinstock. Basic introduction covering isoperimetric problems, theory of elasticity, quantum mechanics, electrostatics, etc. Exercises throughout. 326pp. $5^3/_8$ x $8^1/_2$. 0-486-63069-2

THE CONTINUUM: A CRITICAL EXAMINATION OF THE FOUNDATION OF ANALYSIS, Hermann Weyl. Classic of 20th-century foundational research deals with the conceptual problem posed by the continuum. 156pp. $5^3/_8$ x $8^1/_2$. 0-486-67982-9

CHALLENGING MATHEMATICAL PROBLEMS WITH ELEMENTARY SOLUTIONS, A. M. Yaglom and I. M. Yaglom. Over 170 challenging problems on probability theory, combinatorial analysis, points and lines, topology, convex polygons, many other topics. Solutions. Total of 445pp. $5^3/_8$ x $8^1/_2$. Two-vol. set.
Vol. I: 0-486-65536-9 Vol. II: 0-486-65537-7

INTRODUCTION TO PARTIAL DIFFERENTIAL EQUATIONS WITH APPLICATIONS, E. C. Zachmanoglou and Dale W. Thoe. Essentials of partial differential equations applied to common problems in engineering and the physical sciences. Problems and answers. 416pp. $5^3/_8$ x $8^1/_2$. 0-486-65251-3

STOCHASTIC PROCESSES AND FILTERING THEORY, Andrew H. Jazwinski. This unified treatment presents material previously available only in journals, and in terms accessible to engineering students. Although theory is emphasized, it discusses numerous practical applications as well. 1970 edition. 400pp. $5^3/_8$ x $8^1/_2$. 0-486-46274-9

Math—Decision Theory, Statistics, Probability

INTRODUCTION TO PROBABILITY, John E. Freund. Featured topics include permutations and factorials, probabilities and odds, frequency interpretation, mathematical expectation, decision-making, postulates of probability, rule of elimination, much more. Exercises with some solutions. Summary. 1973 edition. 247pp. 5⅜ x 8½.
0-486-67549-1

STATISTICAL AND INDUCTIVE PROBABILITIES, Hugues Leblanc. This treatment addresses a decades-old dispute among probability theorists, asserting that both statistical and inductive probabilities may be treated as sentence-theoretic measurements, and that the latter qualify as estimates of the former. 1962 edition. 160pp. 5⅜ x 8½.
0-486-44980-7

APPLIED MULTIVARIATE ANALYSIS: Using Bayesian and Frequentist Methods of Inference, Second Edition, S. James Press. This two-part treatment deals with foundations as well as models and applications. Topics include continuous multivariate distributions; regression and analysis of variance; factor analysis and latent structure analysis; and structuring multivariate populations. 1982 edition. 692pp. 5⅜ x 8½.
0-486-44236-5

LINEAR PROGRAMMING AND ECONOMIC ANALYSIS, Robert Dorfman, Paul A. Samuelson and Robert M. Solow. First comprehensive treatment of linear programming in standard economic analysis. Game theory, modern welfare economics, Leontief input-output, more. 525pp. 5⅜ x 8½.
0-486-65491-5

PROBABILITY: AN INTRODUCTION, Samuel Goldberg. Excellent basic text covers set theory, probability theory for finite sample spaces, binomial theorem, much more. 360 problems. Bibliographies. 322pp. 5⅜ x 8½.
0-486-65252-1

GAMES AND DECISIONS: INTRODUCTION AND CRITICAL SURVEY, R. Duncan Luce and Howard Raiffa. Superb nontechnical introduction to game theory, primarily applied to social sciences. Utility theory, zero-sum games, n-person games, decision-making, much more. Bibliography. 509pp. 5⅜ x 8½.
0-486-65943-7

INTRODUCTION TO THE THEORY OF GAMES, J. C. C. McKinsey. This comprehensive overview of the mathematical theory of games illustrates applications to situations involving conflicts of interest, including economic, social, political, and military contexts. Appropriate for advanced undergraduate and graduate courses; advanced calculus a prerequisite. 1952 ed. x+372pp. 5⅜ x 8½.
0-486-42811-7

FIFTY CHALLENGING PROBLEMS IN PROBABILITY WITH SOLUTIONS, Frederick Mosteller. Remarkable puzzlers, graded in difficulty, illustrate elementary and advanced aspects of probability. Detailed solutions. 88pp. 5⅜ x 8½.
0-486-65355-2

PROBABILITY THEORY: A CONCISE COURSE, Y. A. Rozanov. Highly readable, self-contained introduction covers combination of events, dependent events, Bernoulli trials, etc. 148pp. 5⅜ x 8¼.
0-486-63544-9

THE STATISTICAL ANALYSIS OF EXPERIMENTAL DATA, John Mandel. First half of book presents fundamental mathematical definitions, concepts and facts while remaining half deals with statistics primarily as an interpretive tool. Well-written text, numerous worked examples with step-by-step presentation. Includes 116 tables. 448pp. 5⅜ x 8½.
0-486-64666-1

Math—Geometry and Topology

ELEMENTARY CONCEPTS OF TOPOLOGY, Paul Alexandroff. Elegant, intuitive approach to topology from set-theoretic topology to Betti groups; how concepts of topology are useful in math and physics. 25 figures. 57pp. 5⅜ x 8½.   0-486-60747-X

A LONG WAY FROM EUCLID, Constance Reid. Lively guide by a prominent historian focuses on the role of Euclid's Elements in subsequent mathematical developments. Elementary algebra and plane geometry are sole prerequisites. 80 drawings. 1963 edition. 304pp. 5⅜ x 8½.   0-486-43613-6

EXPERIMENTS IN TOPOLOGY, Stephen Barr. Classic, lively explanation of one of the byways of mathematics. Klein bottles, Moebius strips, projective planes, map coloring, problem of the Koenigsberg bridges, much more, described with clarity and wit. 43 figures. 210pp. 5⅜ x 8½.   0-486-25933-1

THE GEOMETRY OF RENÉ DESCARTES, René Descartes. The great work founded analytical geometry. Original French text, Descartes's own diagrams, together with definitive Smith-Latham translation. 244pp. 5⅜ x 8½.   0-486-60068-8

EUCLIDEAN GEOMETRY AND TRANSFORMATIONS, Clayton W. Dodge. This introduction to Euclidean geometry emphasizes transformations, particularly isometries and similarities. Suitable for undergraduate courses, it includes numerous examples, many with detailed answers. 1972 ed. viii+296pp. 6⅛ x 9¼.   0-486-43476-1

EXCURSIONS IN GEOMETRY, C. Stanley Ogilvy. A straightedge, compass, and a little thought are all that's needed to discover the intellectual excitement of geometry. Harmonic division and Apollonian circles, inversive geometry, hexlet, Golden Section, more. 132 illustrations. 192pp. 5⅜ x 8½.   0-486-26530-7

THE THIRTEEN BOOKS OF EUCLID'S ELEMENTS, translated with introduction and commentary by Sir Thomas L. Heath. Definitive edition. Textual and linguistic notes, mathematical analysis. 2,500 years of critical commentary. Unabridged. 1,414pp. 5⅜ x 8½. Three-vol. set.
Vol. I: 0-486-60088-2   Vol. II: 0-486-60089-0   Vol. III: 0-486-60090-4

SPACE AND GEOMETRY: IN THE LIGHT OF PHYSIOLOGICAL, PSYCHOLOGICAL AND PHYSICAL INQUIRY, Ernst Mach. Three essays by an eminent philosopher and scientist explore the nature, origin, and development of our concepts of space, with a distinctness and precision suitable for undergraduate students and other readers. 1906 ed. vi+148pp. 5⅜ x 8½.   0-486-43909-7

GEOMETRY OF COMPLEX NUMBERS, Hans Schwerdtfeger. Illuminating, widely praised book on analytic geometry of circles, the Moebius transformation, and two-dimensional non-Euclidean geometries. 200pp. 5⅝ x 8¼.   0-486-63830-8

DIFFERENTIAL GEOMETRY, Heinrich W. Guggenheimer. Local differential geometry as an application of advanced calculus and linear algebra. Curvature, transformation groups, surfaces, more. Exercises. 62 figures. 378pp. 5⅜ x 8½.   0-486-63433-7

History of Math

THE WORKS OF ARCHIMEDES, Archimedes (T. L. Heath, ed.). Topics include the famous problems of the ratio of the areas of a cylinder and an inscribed sphere; the measurement of a circle; the properties of conoids, spheroids, and spirals; and the quadrature of the parabola. Informative introduction. clxxxvi+326pp. 5³/₈ x 8¹/₂.   0-486-42084-1

A SHORT ACCOUNT OF THE HISTORY OF MATHEMATICS, W. W. Rouse Ball. One of clearest, most authoritative surveys from the Egyptians and Phoenicians through 19th-century figures such as Grassman, Galois, Riemann. Fourth edition. 522pp. 5³/₈ x 8¹/₂.   0-486-20630-0

THE HISTORY OF THE CALCULUS AND ITS CONCEPTUAL DEVELOP-MENT, Carl B. Boyer. Origins in antiquity, medieval contributions, work of Newton, Leibniz, rigorous formulation. Treatment is verbal. 346pp. 5³/₈ x 8¹/₂.   0-486-60509-4

THE HISTORICAL ROOTS OF ELEMENTARY MATHEMATICS, Lucas N. H. Bunt, Phillip S. Jones, and Jack D. Bedient. Fundamental underpinnings of modern arithmetic, algebra, geometry and number systems derived from ancient civilizations. 320pp. 5³/₈ x 8¹/₂.   0-486-25563-8

THE HISTORY OF THE CALCULUS AND ITS CONCEPTUAL DEVELOP-MENT, Carl B. Boyer. Fluent description of the development of both the integral and differential calculus—its early beginnings in antiquity, medieval contributions, and a consideration of Newton and Leibniz. 368pp. 5³/₈ x 8¹/₂.   0-486-60509-4

GAMES, GODS & GAMBLING: A HISTORY OF PROBABILITY AND STATISTICAL IDEAS, F. N. David. Episodes from the lives of Galileo, Fermat, Pascal, and others illustrate this fascinating account of the roots of mathematics. Features thought-provoking references to classics, archaeology, biography, poetry. 1962 edition. 304pp. 5³/₈ x 8¹/₂. (Available in U.S. only.)   0-486-40023-9

OF MEN AND NUMBERS: THE STORY OF THE GREAT MATHEMATICIANS, Jane Muir. Fascinating accounts of the lives and accomplishments of history's greatest mathematical minds—Pythagoras, Descartes, Euler, Pascal, Cantor, many more. Anecdotal, illuminating. 30 diagrams. Bibliography. 256pp. 5³/₈ x 8¹/₂.   0-486-28973-7

HISTORY OF MATHEMATICS, David E. Smith. Nontechnical survey from ancient Greece and Orient to late 19th century; evolution of arithmetic, geometry, trigonometry, calculating devices, algebra, the calculus. 362 illustrations. 1,355pp. 5³/₈ x 8¹/₂. Two-vol. set.   Vol. I: 0-486-20429-4   Vol. II: 0-486-20430-8

A CONCISE HISTORY OF MATHEMATICS, Dirk J. Struik. The best brief history of mathematics. Stresses origins and covers every major figure from ancient Near East to 19th century. 41 illustrations. 195pp. 5³/₈ x 8¹/₂.   0-486-60255-9

Physics

OPTICAL RESONANCE AND TWO-LEVEL ATOMS, L. Allen and J. H. Eberly. Clear, comprehensive introduction to basic principles behind all quantum optical resonance phenomena. 53 illustrations. Preface. Index. 256pp. 5⅜ x 8½.   0-486-65533-4

QUANTUM THEORY, David Bohm. This advanced undergraduate-level text presents the quantum theory in terms of qualitative and imaginative concepts, followed by specific applications worked out in mathematical detail. Preface. Index. 655pp. 5⅜ x 8½.
0-486-65969-0

ATOMIC PHYSICS (8th EDITION), Max Born. Nobel laureate's lucid treatment of kinetic theory of gases, elementary particles, nuclear atom, wave-corpuscles, atomic structure and spectral lines, much more. Over 40 appendices, bibliography. 495pp. 5⅜ x 8½.
0-486-65984-4

A SOPHISTICATE'S PRIMER OF RELATIVITY, P. W. Bridgman. Geared toward readers already acquainted with special relativity, this book transcends the view of theory as a working tool to answer natural questions: What is a frame of reference? What is a "law of nature"? What is the role of the "observer"? Extensive treatment, written in terms accessible to those without a scientific background. 1983 ed. xlviii+172pp. 5⅜ x 8½.
0-486-42549-5

AN INTRODUCTION TO HAMILTONIAN OPTICS, H. A. Buchdahl. Detailed account of the Hamiltonian treatment of aberration theory in geometrical optics. Many classes of optical systems defined in terms of the symmetries they possess. Problems with detailed solutions. 1970 edition. xv + 360pp. 5⅜ x 8½.   0-486-67597-1

PRIMER OF QUANTUM MECHANICS, Marvin Chester. Introductory text examines the classical quantum bead on a track: its state and representations; operator eigenvalues; harmonic oscillator and bound bead in a symmetric force field; and bead in a spherical shell. Other topics include spin, matrices, and the structure of quantum mechanics; the simplest atom; indistinguishable particles; and stationary-state perturbation theory. 1992 ed. xiv+314pp. 6⅛ x 9¼.   0-486-42878-8

LECTURES ON QUANTUM MECHANICS, Paul A. M. Dirac. Four concise, brilliant lectures on mathematical methods in quantum mechanics from Nobel Prize-winning quantum pioneer build on idea of visualizing quantum theory through the use of classical mechanics. 96pp. 5⅜ x 8½.   0-486-41713-1

THIRTY YEARS THAT SHOOK PHYSICS: THE STORY OF QUANTUM THEORY, George Gamow. Lucid, accessible introduction to influential theory of energy and matter. Careful explanations of Dirac's anti-particles, Bohr's model of the atom, much more. 12 plates. Numerous drawings. 240pp. 5⅜ x 8½.   0-486-24895-X

ELECTRONIC STRUCTURE AND THE PROPERTIES OF SOLIDS: THE PHYSICS OF THE CHEMICAL BOND, Walter A. Harrison. Innovative text offers basic understanding of the electronic structure of covalent and ionic solids, simple metals, transition metals and their compounds. Problems. 1980 edition. 582pp. 6⅛ x 9¼.
0-486-66021-4

HYDRODYNAMIC AND HYDROMAGNETIC STABILITY, S. Chandrasekhar. Lucid examination of the Rayleigh-Benard problem; clear coverage of the theory of instabilities causing convection. 704pp. 5³/₈ x 8¹/₄. 0-486-64071-X

INVESTIGATIONS ON THE THEORY OF THE BROWNIAN MOVEMENT, Albert Einstein. Five papers (1905–8) investigating dynamics of Brownian motion and evolving elementary theory. Notes by R. Fürth. 122pp. 5³/₈ x 8¹/₂. 0-486-60304-0

THE PHYSICS OF WAVES, William C. Elmore and Mark A. Heald. Unique overview of classical wave theory. Acoustics, optics, electromagnetic radiation, more. Ideal as classroom text or for self-study. Problems. 477pp. 5³/₈ x 8¹/₂. 0-486-64926-1

GRAVITY, George Gamow. Distinguished physicist and teacher takes reader-friendly look at three scientists whose work unlocked many of the mysteries behind the laws of physics: Galileo, Newton, and Einstein. Most of the book focuses on Newton's ideas, with a concluding chapter on post-Einsteinian speculations concerning the relationship between gravity and other physical phenomena. 160pp. 5³/₈ x 8¹/₂. 0-486-42563-0

PHYSICAL PRINCIPLES OF THE QUANTUM THEORY, Werner Heisenberg. Nobel Laureate discusses quantum theory, uncertainty, wave mechanics, work of Dirac, Schroedinger, Compton, Wilson, Einstein, etc. 184pp. 5³/₈ x 8¹/₂. 0-486-60113-7

ATOMIC SPECTRA AND ATOMIC STRUCTURE, Gerhard Herzberg. One of best introductions; especially for specialist in other fields. Treatment is physical rather than mathematical. 80 illustrations. 257pp. 5³/₈ x 8¹/₂. 0-486-60115-3

AN INTRODUCTION TO STATISTICAL THERMODYNAMICS, Terrell L. Hill. Excellent basic text offers wide-ranging coverage of quantum statistical mechanics, systems of interacting molecules, quantum statistics, more. 523pp. 5³/₈ x 8¹/₂. 0-486-65242-4

THEORETICAL PHYSICS, Georg Joos, with Ira M. Freeman. Classic overview covers essential math, mechanics, electromagnetic theory, thermodynamics, quantum mechanics, nuclear physics, other topics. First paperback edition. xxiii + 885pp. 5³/₈ x 8¹/₂. 0-486-65227-0

PROBLEMS AND SOLUTIONS IN QUANTUM CHEMISTRY AND PHYSICS, Charles S. Johnson, Jr. and Lee G. Pedersen. Unusually varied problems, detailed solutions in coverage of quantum mechanics, wave mechanics, angular momentum, molecular spectroscopy, more. 280 problems plus 139 supplementary exercises. 430pp. 6¹/₂ x 9¹/₄. 0-486-65236-X

THEORETICAL SOLID STATE PHYSICS, Vol. 1: Perfect Lattices in Equilibrium; Vol. II: Non-Equilibrium and Disorder, William Jones and Norman H. March. Monumental reference work covers fundamental theory of equilibrium properties of perfect crystalline solids, non-equilibrium properties, defects and disordered systems. Appendices. Problems. Preface. Diagrams. Index. Bibliography. Total of 1,301pp. 5³/₈ x 8¹/₂. Two volumes. Vol. I: 0-486-65015-4 Vol. II: 0-486-65016-2

WHAT IS RELATIVITY? L. D. Landau and G. B. Rumer. Written by a Nobel Prize physicist and his distinguished colleague, this compelling book explains the special theory of relativity to readers with no scientific background, using such familiar objects as trains, rulers, and clocks. 1960 ed. vi+72pp. 5³/₈ x 8¹/₂. 0-486-42806-0

A TREATISE ON ELECTRICITY AND MAGNETISM, James Clerk Maxwell. Important foundation work of modern physics. Brings to final form Maxwell's theory of electromagnetism and rigorously derives his general equations of field theory. 1,084pp. 5⅜ x 8½. Two-vol. set.   Vol. I: 0-486-60636-8   Vol. II: 0-486-60637-6

MATHEMATICS FOR PHYSICISTS, Philippe Dennery and Andre Krzywicki. Superb text provides math needed to understand today's more advanced topics in physics and engineering. Theory of functions of a complex variable, linear vector spaces, much more. Problems. 1967 edition. 400pp. 6½ x 9¼.   0-486-69193-4

INTRODUCTION TO QUANTUM MECHANICS WITH APPLICATIONS TO CHEMISTRY, Linus Pauling & E. Bright Wilson, Jr. Classic undergraduate text by Nobel Prize winner applies quantum mechanics to chemical and physical problems. Numerous tables and figures enhance the text. Chapter bibliographies. Appendices. Index. 468pp. 5⅜ x 8½.   0-486-64871-0

METHODS OF THERMODYNAMICS, Howard Reiss. Outstanding text focuses on physical technique of thermodynamics, typical problem areas of understanding, and significance and use of thermodynamic potential. 1965 edition. 238pp. 5⅜ x 8½.
0-486-69445-3

THE ELECTROMAGNETIC FIELD, Albert Shadowitz. Comprehensive under- graduate text covers basics of electric and magnetic fields, builds up to electromagnetic theory. Also related topics, including relativity. Over 900 problems. 768pp. 5⅝ x 8¼.
0-486-65660-8

GREAT EXPERIMENTS IN PHYSICS: FIRSTHAND ACCOUNTS FROM GALILEO TO EINSTEIN, Morris H. Shamos (ed.). 25 crucial discoveries: Newton's laws of motion, Chadwick's study of the neutron, Hertz on electromagnetic waves, more. Original accounts clearly annotated. 370pp. 5⅜ x 8½.   0-486-25346-5

EINSTEIN'S LEGACY, Julian Schwinger. A Nobel Laureate relates fascinating story of Einstein and development of relativity theory in well-illustrated, nontechnical volume. Subjects include meaning of time, paradoxes of space travel, gravity and its effect on light, non-Euclidean geometry and curving of space-time, impact of radio astronomy and space-age discoveries, and more. 189 b/w illustrations. xiv+250pp. 8⅜ x 9¼.   0-486-41974-6

THE VARIATIONAL PRINCIPLES OF MECHANICS, Cornelius Lanczos. Philosophic, less formalistic approach to analytical mechanics offers model of clear, scholarly exposition at graduate level with coverage of basics, calculus of variations, principle of virtual work, equations of motion, more. 418pp. 5⅜ x 8½.   0-486-65067-7

Paperbound unless otherwise indicated. Available at your book dealer, online at www.doverpublications.com, or by writing to Dept. GI, Dover Publications, Inc., 31 East 2nd Street, Mineola, NY 11501. For current price information or for free catalogues (please indicate field of interest), write to Dover Publications or log on to www.doverpublications.com and see every Dover book in print. Dover publishes more than 400 books each year on science, elementary and advanced mathematics, biology, music, art, literary history, social sciences, and other areas.

A. Ya. Khinchin
1894–1959